Palgrave Studies in the History of Science and Technology

Designed to bridge the gap between the history of science and the history of technology, this series publishes the best new work by promising and accomplished authors in both areas. In particular, it offers historical perspectives on issues of current and ongoing concern, provides international and global perspectives on scientific issues, and encourages productive communication between historians and practicing scientists.

More information about this series at
http://www.palgrave.com/gp/series/14581

Peter Ayres

Women and the Natural Sciences in Edwardian Britain

In Search of Fellowship

palgrave
macmillan

Peter Ayres
Lancaster University
Lancaster, UK

Palgrave Studies in the History of Science and Technology
ISBN 978-3-030-46599-5 ISBN 978-3-030-46600-8 (eBook)
https://doi.org/10.1007/978-3-030-46600-8

Cover illustration: World History Archive / Alamy Stock Photo

This Palgrave Macmillan imprint is published by the registered company Springer Nature Switzerland AG.
The registered company address is: Gewerbestrasse 11, 6330 Cham, Switzerland

Preface: 'Associations of Persons United by Some Common Interest'

Through the Victorian era, age-old prejudices still prevailed concerning the fitness of women to be scientists or, more generally, to enter the professions. It was pointed out by men that the weight of a woman's brain was less than that of a man's; others argued that the physical exertion of a working life would imperil a woman's reproductive health (Cock and Forsdyke 2008, 178; Rayner-Canham and Rayner-Canham 2003). Or, quite simply, it was held that a woman's first duty was to support her husband and his children, and not to spend her time pursuing some high-flown science. Such attitudes infected even the thinking of women; writing about 'Our School Girls', Mrs CE Humphry, one of the first female journalists (and an extremely popular one) reflected;

> *What is the use of class successes if they are won at the expense of health? And though scholarships are very pleasant things…they may cost too dear. If the money they save has to go on doctors' fees, of what earthly use are they.* (Humphry 1898, 19)

It is neither the purpose of this book to examine how those particular prejudices were overcome, nor to review the debate that 'the mind has no sex'—a debate which has stretched down the years, from Lydia Becker's proposal to a meeting of the British Association for the Advancement of Science (1868) that differences between the minds of women and men were a result of nurture rather than nature, until today (Gianquitto 2013). It is a debate which continues to occupy the energies of some of the finest feminist scholars, who ask whether women 'do' science in a way that is different from the way it is done by men, and, by extension, whether feminism has changed science.

It is the purpose of this book to look at one little explored aspect of the wider debate, the historical exclusion of women from leading scientific societies and the impact that that had on their efforts to become integrated into the world of professional science. The promise held out by the fundamental tenet of 18th century European Enlightenment that 'all men are by nature equal' was not realised for, as the nineteenth century progressed, women were progressively excluded as the culture of science was gradually closed to them (Schiebinger 1999, 13 and 69). There occurred, in parallel, a professionalization of science and a privatisation of the family, the two spheres being, respectively, the domain of men and women.

My own past interests have centred on the professionalization of botany in the decades immediately preceding World War I. In writing about some of the leading men of the time I have been struck by how often their researches were assisted by women, although each for only a short time—suggesting that either a lack of funding, or marriage, ended each woman's connection with her successful man and, thereby, her potential career. The names of a few women do, however, recur again and again in the pages of fledgling journals such as the *New Phytologist* (in which I declare a personal interest) and the *Annals of Botany*. Perhaps they were women who, exceptionally, found permanent employment, or who enjoyed private means? My enquiries into the lives of these women led me to seek comparisons with the lives of women in other natural sciences and, almost inevitably, parallels became apparent—not least the difficulties all women had in acquiring fellowships in scientific societies.

The wider background to the late-Victorian and Edwardian periods involves, of course, women's fight for the right to vote (a fight which itself spawned numerous clubs, associations, and societies). It is not a coincidence that the two struggles were contemporaneous, and as individual lives are explored, it will be seen that the same women were often involved.

The extent to which women in the natural sciences depended on male help is explored, as is the question why some men chose to be 'enablers', when others stood in the way of women's progress. In such analysis, two things should be borne in mind. First, it is probable that the majority of male scientists had no strong views one way or the other. Second, many men were educated in an all-male environment; they knew little about the abilities and interests of females of the same age. Unfamiliarity could all too easily lead to an awkwardness and shyness, resulting in them avoiding social or professional interactions with the opposite sex.

Finally, with tongue somewhat in cheek, I return to my own earlier interests. Writing about Charles Darwin led me to his grandson, Bernard, golf correspondent of *The Times* newspaper from 1907 to 1953. Bernard's interests stretched, however, beyond golf clubs to 'Gentlemen's Clubs', which he called, '...associations of persons united by some common interest meeting periodically for cooperation or conviviality' (Darwin 1943). He could easily have extended his definition to include Societies.

In 1941, with most of Europe under Nazi domination, the London publisher Collins launched a series of social history books called 'Britain in Pictures'. The slim volumes were designed to boost morale but also to record a British way of life that was at risk of extinction. Bernard Darwin was invited to contribute a book on the subject of 'British Clubs' (Darwin 1943). Gentleman's clubs, ranging from dining to debating to sporting ones, and mostly dating from the eighteenth or nineteenth centuries, were, he argued, a defining characteristic of British society. While some of the 'Social Clubs' have since Darwin's time admitted women, for example, The Athenaeum in 2002, others, such as Boodle's, Brooks's, and White's are still for men only. To be fair, the University Women's Club (founded in 1921 as the University Club for Ladies) excludes men, but the overall conclusion is that in London's clubland, centred in St James', at the heart of the metropolis, old habits die hard.

Is it any wonder then that the oldest of Britain's scientific societies, which were effectively gentlemen's clubs, were so resistant to change, so averse to opening their doors to women? Male scientists—supposedly enlightened and rational—were no better than their non-scientific peers.

This book explores how prejudice and ignorance in those societies were slowly overcome by a small band of women, and their sympathetic male supporters. Or, as Bernard might have put it, '...how women became clubbable'.

Lancaster, UK — Peter Ayres

References

Cock, A., and D.R. Forsdyke. 2008. *Treasure Your Exceptions: The Science and Life of William Bateson*. New York: Springer.

Darwin, B. 1943. *British Clubs*. London: William Collins.

Gianquitto, Tina. 2013. Botanical Smuts and Hermaphrodites; Lydia Becker, Darwin's Botany and Education Reform. *Isis* 104: 250–277.

Humphry, Charlotte E. 1898. *A Word to Women*. London: James Bowden.

Rayner-Canham, Marelene F., and G.W. Rayner-Canham. 2003. Pounding on the Doors: The Fight for Acceptance of British Women Chemists. *Bulletin for the History of Chemistry* 28: 110–119.

Schiebinger, Londa. 1999. *Has Feminism Changed Science?* Cambridge, MA: Harvard University Press.

By the Same Author

Harry Marshall Ward and the Fungal Thread of Death. 2005. St. Paul Mn.: American Phytopathological Society.

The Aliveness of Plants. The Darwins at the Dawn of Plant Science. 2008. London: Pickering and Chatto.

Shaping Ecology: The Life of Arthur Tansley. 2012. Chichester: John Wiley.

Medicinal Plants in Wartime. Britain's Green Allies. 2015. Kibworth: Matador.

Acknowledgements

In pursuing the women who sought fellowship I needed practical help from many, many groups and institutions, and it was always given cheerfully and generously. The archivists and historians to whom I owe thanks are Nicola Allen (Woburn Abbey), Gill Butterfill and Philippa Lewis (Kew), Mark Carine, Helen Pethers, and Laura Brown (Natural History Museum), Elen Curran (James Allen's Girls School), Katrina Dean (Cambridge University Library), Brent Elliot (Royal Horticultural Society), William George (Essex Field Club), Alison Harvey (Cardiff University), Mary Henderson (Dundee Women's Trail), Debbie Hunt (Royal Microscopical Society), Nancy Janda (Hunt Institute for Botanical Documentation, Carnegie Mellon University, PA), Pete Kinnear (Dundee University), Paula Lightfoot (Yorkshire Naturalists Union), Gillian Murphy (London School of Economics), Kate O'Donnell (Somerville College, Oxford), Norman Porrett (British Mycological Society), Carol Sandford (Holmesdale Natural History Club), Carol Stewart (Strathclyde University), Naomi Sturges and Matilda Watson (Girton College, Cambridge), Anne Thomson (Newnham College, Cambridge), Annabel Valentine and Ellis Huddart (Royal Holloway College), Hannah Westall (Girton College, Cambridge), Alison Wheatley (Handsworth College). At the Linnean Society, Gina Douglas, and Liz McGow gave invaluable help.

On the trail of Mrs Farquharson, Claire Jones, Sarah Pedersen, Val Pollitt, and Lindy Moore were most helpful; on that of Emma Turner, James Parry kindly supplied a lot of information; and in my pursuit of Lilian Clarke, Dawn Sanders and Roy Vickery gave both insights and encouragement. Cynthia Burek guided me most helpfully on the subject of Maria Ogilvie-Gordon, and John Maris introduced me to the family tree of the Embletons.

I thank Patricia Fara (Churchill College, Cambridge) and Jane Robinson (Somerville College, Oxford) for their interest and support. Old friends, Jeff Duckett and Miriam David, offered much-needed help and advice.

And, finally, the project would never have been initiated or completed without the unwavering support of my wife, Mary. To her I owe my greatest thanks.

Praise for *Women and the Natural Sciences in Edwardian Britain*

"In this compelling history with modern relevance, Peter Ayres describes the female pioneers who realised that scientific success lay in solidarity. Networking towards the future, they campaigned for entry into universities, societies and laboratories, collectively achieving the individual recognition they deserved."

—Patricia Fara, Fellow of Clare College, Cambridge and author of *A Lab of One's Own: Science and Suffrage in the First World War*

"This is discovery in the purest sense of the word: the revelation of something always there, but forgotten. In *Women and the Natural Sciences* Peter Ayres presents us with a group of fascinating pioneers. He polishes away the accretion of convention, institutional prejudice and natural diffidence, allowing their extraordinary—and ordinary—achievements to shine at last."

—Jane Robinson, social historian and author of *Ladies Can't Climb Ladders: the Pioneering Adventures of the First Professional Women*

Contents

List of Figures

CHAPTER 1

Fellowship and a Woman's Place in Edwardian Britain

As the twentieth century opened, women were increasingly challenging a world designed by and for men, their confidence enhanced by the better education they were enjoying. Educational reform, in particular the formation of the Girls Public Day School Company (1872), had led to the foundation of schools that recognised the importance of both the quality of their teaching and the range of subjects taught. Ever greater numbers of girls from upper and middle-class homes were attending school, rather than being educated at home, most girls receiving thereby at least a rudimentary education in the natural sciences. And for many girls, they found science was to their liking. Conveniently for them, educational reforms in late Victorian times had extended to the universities where, in conjunction with the opening of new colleges and halls of residence for women, more science courses were admitting women. Male tutors may not have always been welcoming, limited laboratory facilities were not always shared equally with male students, and field work presented for women special problems associated with dress and chaperonage, but women were not deterred; this in spite of the fact that the many who studied at Oxford or Cambridge were not allowed *formally* to graduate until 1920 and 1947, respectively.

The difficulties experienced by women while undergraduates were nothing compared with those faced subsequently if they wished to undertake post-graduate work and, ultimately, make a career in science. As Marsha Richmond (1997) concluded from her examination of Cambridge's

P. Ayres, *Women and the Natural Sciences in Edwardian Britain*, Palgrave Studies in the History of Science and Technology, https://doi.org/10.1007/978-3-030-46600-8_1

Balfour Biological Laboratory for Women, 'women were excluded from the social community of science'. Unlike The Balfour, few laboratories in Britain offered either bench space or employment for women graduates. A small handful of women were wealthy enough to be able to finance their own research laboratories, while others were able and willing to survive on unpaid work, if they could find laboratory space and a sympathetic research director or head of laboratory. Many more women could only pursue a career in science if they could find paid employment, and that brought them into direct competition with men.

Women's social exclusion from the community of science was due to many factors, not least the contemporary prejudices of many male scientists concerning both the intellectual and physical abilities of women. One aspect of social exclusion, which has remained largely unexplored until now, was the difficulty women faced in joining scientific and learned societies—a difficulty which was a consequence of male prejudices, and a desire for exclusivity. In order to know and be known by potential research directors and employers, a women needed interactions with male scientists of seniority and influence, but how and where could those interactions occur in a proper and socially acceptable manner? The most practical place would be within the learned societies associated with each science. These gave their male members the chance to air their ideas, to test the results of their research, and a means of becoming known personally by their peers, but women were denied those same opportunities because they were denied formal Fellowship of most societies—they were disadvantaged. The botanist and suffrage campaigner, Lydia Becker, argued that such exclusion lay at the heart of 'the scientific disabilities of women' (Bernstein 2006, 87).

This book tells how women successfully fought to be included in the social community of science; specifically, how they won the right to join scientific societies and no longer be disadvantaged as they sought to find a work place and build a career. Success was in some cases attributable to the efforts of individual women, in other cases to the supportive networks which women built. It will be seen that there was support too from sympathetic men; men who often worked within societies to overcome the prejudices of the fellows and persuade them of the advantages of admitting women.

Targets for Middle-Class Women

The term 'Edwardian Era' includes strictly the years, 1901–1910, when Britain was ruled by Edward VII, but it is often stretched, as here, to include the 1890s when, as Prince of Wales, 'Bertie' set the tone of the nation. Both before and after his coronation, in 1902, he openly enjoyed a string of mistresses, Frances 'Daisy' Greville, the Countess of Warwick, being one of them (Heffer 2017, 89). Renowned for her beauty, Daisy was exceptional in another way, for she was a social reformer intent on improving the lot of women—though, for her, this meant those middle-class ones having some education.

According to the national census of 1901, such middle-class women comprised about 5% of that part of adult female population which was self-supporting, either by necessity or choice. The remaining 95% of self-supporting women were from the working-classes, labouring mostly in industry or domestic service and having little or no education. Daisy's particular interest was in agriculture and horticulture and it was in those areas, which she termed 'the lighter classes of agriculture', that she sought to provide training and job placements, healthy alternatives to a dreary life that might otherwise be spent as a governess. The typical target of her plans would be a middle-class women who had a small inheritance but who needed to make it work for her financially, just as it might in a small, well run, horticultural establishment (Scott 2017, 47). In 1898, Daisy established the Lady Warwick Hall (of residence) in Reading, a forerunner of the University of Reading, where women could be taught by staff of the Oxford University Extension College. In 1903 her establishment moved when she set up the much larger and independent Studley Horticultural and Agricultural College for Women, in Warwickshire. Subjects such as entomology found their way onto the curriculum but, generally, there was little emphasis on science *per se*. In dealing with only the 'lighter classes of agriculture', the ambitions of the college were strictly gendered, not extending to full equality of the sexes (Opitz 2014).

In another sphere, however, Daisy was more ambitious. She founded the Lady Warwick Agricultural Association for Women which, as reported by *The Times* of 21st October, 1899, had two days earlier held its first annual meeting—at Stafford House, St James, London—when, as part of the proceedings

> *The chairman moved, and Mrs Garrett Anderson MD seconded, a resolution: That it is desirable and important that duly qualified women should have the advantage of full fellowship in Scientific and other Learned Societies, e.g. the Royal, the Linnean and the Royal Microscopical.*[1]

The targets had been identified.

In support of the motion, a paper by Mrs Farquharson of Haughton was read, though in her absence by Mr R. Moran. Already committed to joining scientific societies, as and when an opportunity arose, Marian Farquharson was greatly encouraged by the tone of Lady Warwick's meeting, its aristocratic leadership, and the publicity it received (Anon. 1899). She was already a member of the Royal Microscopical Society, though not a *full* member since women's involvement with that Society's activities was limited until 1909, and her scientific achievements fell way short of those required for a fellowship of the Royal Society, so she focussed her attention on the Linnean Society, the world's oldest extant biological society. By the end of 1904 she had successfully persuaded that Society to make women fellows—known at the time as 'Lady-Fellows' and sometimes referred to here as Linnaeus' Ladies—although she paid a price, for her own application was rejected. The Linnean was not the first, but it was *among* the first scientific societies to admit women; thanks to the widespread respect in which it was held, it set an important precedent for other societies.

Fellowship and Women

Fellowship, or membership, of such a society was of fundamental importance because it provided not only a vital meeting place where women could, in theory, meet and mix freely with male fellows but it also carried with it a range of other practical benefits. Thus, fellowship gave access to specialist libraries, to museums, and to reference collections, that is to established learning. Fellowship offered places where a passionate interest could be shared with other enthusiasts, and it provided opportunities to learn from friendly experts.

[1] The chair was Mr Marshall Dugdale, barrister and High Sheriff of Montgomeryshire. Stafford House was a home of the 4th Duke of Sutherland. His wife, Millicent, was half-sister to Daisy Warwick and, like her, a renowned society hostess and social reformer.

The idea of fellowship represented something else, more tantalising than even those practical benefits. The very exclusive and elusive nature of fellowship made it a prize in its own right, something which demonstrated women's equality with men.

No longer satisfied with membership of one or more of the various field clubs which had opened up across Britain during the later decades of the nineteenth century, many women were thus actively seeking fellowship, knocking with increasing fervour on the doors of scientific societies, hoping to gain admission—though in many cases being disappointed.

Later chapters will explore the lives of those women who were successful in becoming the Linnean's first female fellows because they provide a panoramic snapshot of women's involvement in the natural sciences in the Edwardian Era, their interests ranging through botany, geology, and genetics, and their qualifications from nothing formal to the possession of higher degrees. Some were the products of the old methods of private tuition while, in contrast, others had passed through well-endowed schools offering a diversity of educational experiences. These, and the women who struggled to join comparable societies, illustrate also how limited was the range of opportunities available in later life, even for those who were the most highly educated and motivated to play an active role in the natural sciences. They were not in the main the highest-fliers scientifically; they were not immortalised by discoveries forever associated with their names. They did, however, commit their lives to the natural sciences, in some cases being paid for their work, in other cases not. By their example they made easier the path for succeeding generations of women who aspired to play a full part in the natural sciences—as the equals of men.

Professionalization of the Natural Sciences

Within late-Victorian Britain a gradual change was happening which was to have a significant bearing upon women's struggle to join scientific societies. It was professionalisation, and it affected the ambitions of men as well as those of women. The source of new knowledge was increasingly the laboratory, a place where studies relied on complex and expensive equipment; equipment operated by highly skilled professionals. Amateurs could still contribute to the sciences, as the lives of many of our subjects will show, but the pressure on women to join the culture of the professional laboratory was growing inexorably. In this respect, the Edwardian era was one of accelerating transition.

The history of British science in the eighteenth and nineteenth centuries is replete with examples of major advances made in country houses, either by the aristocratic or wealthy owners, or by their poorer protégées. Thus, in the late eighteenth century Lord Shelburne's Bowood House in the deepest countryside of Wiltshire became famous not only as a weekend meeting place for the leading intellectuals and politicians of the day but as somewhere that sheltered and provided a laboratory for the researches of two brilliant mavericks, the Unitarian preacher Joseph Priestley and the Dutch émigré Jan Ingen Housz, men who *inter alia* contributed significantly to our understanding of photosynthesis in green plants (Beale and Beale 2011, 411).[2]

Moving forwards 100 years, and involving some whose names will recur later, Lord Rayleigh's scientific endeavours and social circle were based on three large estates; his own, Terling Place near Chelmsford in Essex (where in the West Wing in 1894 he conducted his Nobel Prize winning researches on argon—the 'noble gas' he discovered in collaboration with William Ramsay), Whittingehame in East Lothian (the 86-roomed neo-classical home of his brother-in-law and future Prime Minister, Arthur James Balfour), and Hatfield House, in Hertfordshire. The number of scientific papers produced by Rayleigh from those houses far exceeded those originating from his time (1879–1885) as Director of the Cavendish Laboratory in Cambridge. At Whittingehame there was an extensive collection of fish, bird, insect, and fossil specimens, which helped inspire the career of Arthur's young brother, Francis Maitland Balfour, the future Cambridge zoologist (Chap. 3). The collection of Lepidoptera was especially fine, thanks to their sister Alice's lifelong efforts (she became in 1916 a Fellow of Royal Entomological Society, a society whose doors had always been open to women) (Opitz 2004).

The tensions current through the Edwardian Era are illustrated by the life of Dukinfield Henry Scott, Botanical Secretary of the Linnean Society from 1902 to 1908. On the one hand, Scott was old fashioned for his family wealth meant that he never had to rely on paid employment. On the other hand, he was a thoroughly modern laboratory researcher: like many other British botanists and zoologists, and chemists too, when he was young he had been attracted to study in a German university. 'The chief characteristic of German university life', said Scott, 'was the dominance of research over mere learning' (Scott 1925). There was emphasis on

[2] Ingen Housz was also a guest of the Earl of Warwick, a forebear of Daisy's husband.

laboratory-based empiricism, evidence-based knowledge. Internationally, the sciences were becoming the property of the professional, someone who worked in a laboratory of a university, museum, or similar institution.

A so-called 'New Botany' was brought back to Britain by Scott and his contemporaries, such as Sydney Vines and Harry Marshall Ward, whose attitudes typified a cadre of young botanists whose work would focus on evidence gained through laboratory experiment and who would lead botany into the twentieth century (Ayres 2005, 37). Julius von Sachs, an inspiration for the many young British botanists who visited his laboratories in Würtzburg, deplored the 'country house' style of Charles Darwin (De Chadarevian 1996). Francis Darwin, Charles' son, and Marshall Ward illustrate perfectly the changes affecting the natural sciences; after these two friends studied in Germany, they became colleagues in Cambridge University's Botany School but, whereas Francis' family wealth allowed him to work purely for pleasure, Ward needed every penny of his income because he was the son of an impoverished music teacher from Nottingham. Men like Francis Darwin, and D.H. Scott, were old-style, their work was their hobby; Marshall Ward was the new model, a professional through and through.

As part of the arrival of New Botany in Britain, The Jodrell Laboratory at Kew, opened in 1876 (and, as will be seen later, became a 'home' for a number of the women whose lives are followed here). Joseph Hooker, the Director of the Royal Botanic Gardens, and Charles Darwin's great friend, conceded that 'it was for younger men to take up the experimental approaches… [young men who were] unmistakably scientists and proud of the title' (Endersby 2008, 311).[3] Inevitably, this new approach hastened the eclipse of the amateur who had no access to laboratories or their sophisticated and expensive equipment.

Although women in the Edwardian Era benefited from improving educational opportunities in schools and colleges, the professionalisation that was occurring in the natural sciences served only to disadvantage them further. They found their access to laboratories was hampered by a lack of bursaries and post-graduate scholarships that might enable them to bridge the gap between graduation and permanent employment. And when a salaried position did become available the chances of it being given to a woman were hampered by all-male appointing committees who rarely

[3] The term 'scientist' was still relatively new, having been coined by William Whewell in 1833. The term most widely used previously had been 'natural philosopher'.

gave female applicants serious consideration. All too often women had to settle for an amateur's life, with little or no access to the tools they needed if they were to contribute to contemporary research.

The integration of women into the mainstream of the natural sciences is one aspect of their wider struggle for equal rights and opportunities. Evolution towards equality in society was not *at the time* seen as inevitable and was resisted and resented by large swathes of the male population—and by some women. The difficulties faced by women in persuading societies such as the Linnean to admit them as fellows should not be underestimated, nor should be the achievements of those women who, once admitted, enriched those societies. It should be no surprise then that women of such energy and determination were often involved also in the suffrage movement.

Demonstrating Their Worth

'Votes for women' was a cause whose profile and strength had been growing ever since the Great Reform Act of 1832 had significantly widened male suffrage. In 1866 John Stuart Mill MP had presented in parliament a petition for women's suffrage bearing 1500 signatures collected by Barbara Bodichon and the Women's Suffrage Committee. Their petition was rejected. Further Reform Bills had by 1884 enfranchised every man paying an annual rental of £10 or holding land valued at £10, but still all women were excluded. To add to the woes of a woman who chose to marry, until the 1882 Act giving women rights of ownership, all her wealth and possessions became the property of her husband upon marriage.[4]

The 1880s saw the birth of various groups which vociferously campaigned for women's suffrage. In 1897 the two largest of these groups combined to form the National Union of Women's Suffrage Societies (NUWSS, its members known as the 'suffragists', their President Millicent Garrett Fawcett) but it was not until 1903 and the formation by Emmeline Pankhurst of the more militant Women's Social and Political Union (WSPU, the 'suffragettes', advocating deeds not words) that suffrage approached the top of the political agenda. Emmeline, and her barrister husband before her, had been inspired by the fiery speeches of their fellow Mancunian, Lydia Becker. By the outbreak of World War I (WWI), female fellows of scientific societies had already proved they could play a full part

[4] Married Women's Property Act, 1882.

in the life of those societies to which they had been admitted. Women were not granted the right to vote in parliamentary elections in Great Britain until 1918; even then they had to be householders and over the age of thirty. Nevertheless, a dam had been breached in 1918; ten years later all women over the age of twenty-one gained the right to vote.[5]

It was not just the many jobs which women had smoothly taken over from men during World War I (many of which were taken back by men as soon as the war was over) that had finally persuaded politicians to grant women the vote but it was also the many small advances which women had been making in the decades immediately preceding the war, advances which included their greater participation in, and contribution to, the sciences. As Patricia Fara, a past-President of the British Association for the History of Science, recently put it in *A Lab of One's Own*, 'Rather than reversing entrenched attitudes overnight, the War made earlier shifts apparent and enabled change to continue—it revealed and accelerated processes of transformation that had begun previously' (Fara 2018, 27–28).

References

Anonymous. 1899. Notes. *Nature*, 26 October, 60, 621–625.

Ayres, P.G. 2005. *Harry Marshall Ward and the Fungal Thread of Death.* St. Paul, MN: American Phytopathological Society.

Beale, N., and Elaine Beale. 2011. *Echoes of Ingen Housz. The Long Lost Story of the Genius Who Rescued the Habsburgs from Smallpox and Became the Father of Photosynthesis.* Salisbury: Hobnob Press.

Bernstein, Susan D. 2006. 'Supposed Differences': Lydia Becker and Victorian Women's Participation in the BAAS. In *Repositioning the Victorian Sciences: Shifting Centres in Nineteenth-Century Scientific Thinking*, ed. David Clifford, Elisabeth Wadge, Alex Warwick, and Martin Willis, 85–93. New York: Anthem.

De Chadarevian, Soraya. 1996. Laboratory Science Versus Country-House Experiments. The Controversy between Julius Sachs and Charles Darwin. *British Journal of the History of Science* 29: 17–41.

Endersby, J. 2008. *Imperial Nature. Joseph Hooker and the Practices of Victorian Science.* Chicago: University Press.

Fara, Patricia. 2018. *A Lab of One's Own. Science and Suffrage in the First World War.* Oxford: Oxford University Press.

Heffer, S. 2017. *The Age of Decadence. Britain 1880 to 1914.* London: Random House Books.

[5] Representation of the People Acts, 1918 and 1928.

Opitz, D.L. 2004. 'Behind Folding Shutters in Whittingehame House': Alice Blanche Balfour (1850–1936) and Amateur Natural History. *Archives of Natural History* 31: 330–384.

———. 2014. 'Back to the Land': Lady Warwick and the Movement for Women's Collegiate Agricultural Education. *Agricultural History Review* 62: 119–145.

Richmond, Marsha L. 1997. 'A Lab of One's Own'. The Balfour Biological Laboratory for Women at Cambridge University, 1884–1914. *Isis* 88: 422–455.

Scott, D.H. 1925. German Reminiscences of the Early Eighties. *New Phytologist* 24: 9–16.

Scott, Caroline. 2017. *Holding the Home Front. The Women's Land Army in the First World War*. Barnsley: Pen & Sword Press.

CHAPTER 2

Joining the Like-Minded. Societies and Meeting Places

They [scientific societies] draw an arbitrary line among scientific students and say to one half of the human race—you shall not enter into the advantages we have to offer.
—Lydia Becker to The Manchester Ladies' Literary Society, 30 January 1867 (cited by Parker 2001, 633)

The first annual conference of Lady Warwick's Association (1899) was a newsworthy event, reported not just in *The Times* newspaper but also in the journal *Nature* where the anonymous correspondent, having reported the motion that women should be able to join scientific societies, reviewed recent moves in that direction;

> *Six years ago the Council of the Royal Geographical Society elected several ladies as ordinary fellows, but their action was disapproved at two special meetings, and resolutions to the effect that it was inexpedient to admit ladies as ordinary fellows were carried by conclusive votes. Ladies are however admitted to meetings of the Society, and papers are accepted from them. In the case of the Royal Astronomical Society, ladies are only admitted to the ordinary meetings by special invitation of the president, sanctioned by the Council.*

The article concluded discouragingly:

P. Ayres, *Women and the Natural Sciences in Edwardian Britain*,
Palgrave Studies in the History of Science and Technology,
https://doi.org/10.1007/978-3-030-46600-8_2

> ...[overall] *there is distinct opposition to the admittance of women at present, and no sudden change of feeling can be expected.* (Anon. 1899)

A broader sweep of women's status in scientific societies at that time presents a mixed picture though one which, on balance, confirms the negative mood. The atmosphere can only have been extremely discouraging for women who sought fellowship with like-minded men and women. What follows underlines how remarkable it was then that only five years after the *Nature* article the Linnean Society agreed to admit women as full members.

Scientific and Learned Societies

Lady Warwick's Association had resolved that women should have full fellowship in scientific societies, and at the top of its list of targets was The Royal Society of London, founded in 1660 and blessed by King Charles II. The Royal was at the top because it was, as it remains, the most prestigious scientific society in Britain. An historic challenge to the Royal came in 1902 with the candidature of a married woman, Hertha Ayrton (née Marks), a candidature supported by nine fellows. Lawyers were consulted. Their opinion was that

> *Whether the [Royal] Charters admit of the election of unmarried women appears to us to be very doubtful. ...A woman, if elected, would become disqualified by marriage.* (Mason 1991, 1995)

Disqualification would occur because in common law a woman's person was covered by that of her father before marriage, and by her husband after marriage. Only in 1919 did women finally and unequivocally become legal 'persons' in their own right following the passage of the Sex Disqualification (Removal) Act through parliament.

The officers of the Royal Society were in no doubt that Mrs Ayrton was ineligible. She did however become in 1904 the first woman to read one of her own papers before a meeting of the Society. Ayrton was in 1906 only the second woman to be honoured with the award of the Society's esteemed Hughes medal (the first being her good friend Marie Curie) for her outstanding work on the electric arc and sand ripples, prompting the Royal's President, William Huggins, to reflect grudgingly, 'Can we now refuse the Fellowship to a medallist?' (Mason 1991). If there had been serious hopes that the attitude of the Royal Society was softening, that a breakthrough was imminent, they were to be disappointed. It was not until 1945 that the first woman, Kathleen Lonsdale, a physicist, was made a Fellow of the Royal Society. She was closely followed in the same year by Marjory Stephenson (microbial biochemistry) and in 1946 by Agnes Arber, a botanist mentored in her youth by Ethel Sargant (Chap. 5).

History is silent about Hertha's response to the failed Royal Society nomination; rejection can only have hardened her resolve to fight for equality, a fight which she carried onto the streets of London as she marched with Emmeline Pankhurst and her suffragettes (Mason 2004).

The Royal Astronomical Society (RAS, founded 1820) similarly relied on its Charter and Statutes, in which there was mention only of 'he', not 'she', to justify its resistance to having female fellows, admitting the first four only in 1916. This was in spite of some clever sophistry whereby as early as 1835 it had found itself able to give Honorary Membership to the exceptionally distinguished Caroline Herschel and Mary Somerville (Bailey 2016). Until 1916, female astronomers had to be content with the membership of either a local society (the first enduring society was Liverpool's, founded in 1881), or, after 1890 and thanks largely to the energy and drive of Elizabeth Brown, the British Astronomical Association (BAA) (Chapman 2016). The local societies often published journals and other documents promoting 'popular' astronomy, for that science was, like botany and geology, one which attracted large numbers of amateurs. However, while these amateurs could take pleasure from their studies of the heavens, advances in knowledge were increasingly being made by professionals, men who had access to ever more powerful and expensive telescopes, and who often shunned the BAA (Chapman 2016; Meadows 2008, 223).

In sharp contrast to the Royal Society and the RAS, in some other societies, such as the Physical Society of London (founded 1874, incorporated into the Institute of Physics in 1921), women were welcome from their inception. As was typical of the time, the Physical Society was open to both amateurs and professionals but, unusually, new members gained immediate access to laboratories—those of the Royal College of Science, South Kensington—where the society's meetings were held. The society chose as its president for 1890–1892 Hertha's husband, William Edward Ayrton.

In some even older societies, such as the Zoological Society (founded 1829) or the Royal Entomological Society (1833), similarly, there had never been discrimination against women. The Royal Horticulture Society (RHS), founded 1804, discovered after a quarter of a century that its rules did not actually preclude women from fellowships, and awarded the first of them in 1830. Choosing to add lustre to itself, the RHS made Lady Radnor the first recipient of a fellowship, just as in later years the Linnean would choose a member of the aristocracy, the Duchess of Bedford (*aka* Mary Russell), to be among its first women fellows. Societies often offered a fudge, as in the case of the Royal Microscopical Society, which boasted of four female members by 1884 but did not until 1909 allow women to attend its meetings, or to take part in its discussions.

The Physiological Society was founded in 1876 as a dining club. Although it never explicitly excluded women, and occasionally they contributed to its meetings and its journals (thirteen women, including Alice L. Embleton, see Chap. 6, contributed full papers to the *Quarterly Journal of Experimental Physiology* between 1908 and 1915), none was ever inducted as a member (Tansey 2015).[1] When in 1912 the question of female membership at last come to the fore, the Society displayed perhaps the most bizarre reason ever for excluding a woman when it argued that, 'it would be improper to dine with ladies [while] smelling of dogs'. The rationale was that the Society's scientific proceedings, which often involved practical demonstrations using dogs or other live animals, routinely preceded the dinners.

As with so many societies, there was no quick resolution of the female problem and it was not until 1915 that a formal resolution was finally agreed at the AGM allowing women to become members. Six were admitted in the first tranche, the first on the alphabetical list, and probably the most distinguished, being Florence Buchanan. After graduating from UCL, she had done post-graduate research under the supervision of Edwin Ray Lankester before moving in 1894 to Oxford to work as research assistant to the Regius Professor of Medicine, John Burdon Sanderson. When Sanderson retired in 1904, she stayed in Oxford working with his nephew, the physiologist, John Scott Haldane. It was Haldane more than anyone who was responsible for helping Florence and the other women to become members of the Physiological Society.[2]

[1] Among authors of shorter communications in the same period were Florence Durham and Edith Rebecca Saunders (Chap. 8).

[2] Haldane's schoolboy friends in Edinburgh included William Herdmann (Chap. 2) and D'Arcy Thompson (Chap. 7); together, they went botanising and hunting fossils (Goodman 2007, 55–56).

Geology, like botany, occupies a special position in the natural sciences because it has always attracted large numbers of amateurs and, as will be seen in the next chapter, geology was in the eighteenth and nineteenth centuries regarded as a subject especially suitable for study by ladies. In spite of this, the Geological Society (founded 1807) battled long and hard against the idea of having female fellows, Thomas Henry Huxley being one of those who spoke against women's membership. Another heavyweight of the geological world, Charles Lyell, was of the opposite mind set; with irony, he likened Huxley's pre-diluvian attitude to that of his old opponent, the Bishop of Oxford, in the infamous evolution debate held at the 1860 meeting of the British Association for the Advancement of Science in Oxford (Desmond 1994, 272–273).

Women were allowed to publish in the Society's journals, and to have their papers read at its meetings by male fellows—strictly in the woman's absence—but women themselves were denied fellowship, an issue which through the 1880s and 1890s was discussed more and more frequently and with increasing passion by its male fellows. It was not until 21 March 1901, however, that the first positive, or, more accurately, provocative action was taken. On that day, Sir Archibald Geikie, a keen supporter of women's equality, brought a Mrs Marian Farquharson and a Miss NM Stewart, neither of them geologists, to an Ordinary General Meeting of the Geological Society. Provocative because it was only twelve months earlier that the Society's Council had had a motion put before it that

> *It is not desirable that Fellows of the Society should be allowed to introduce ladies at the Ordinary General Meetings.* (Cited by Herries Davies 2007, 160)

Although that motion had been defeated, the temperature of the debate between pro- and anti-women factions had been raised significantly. The pro-camp won a partial victory in 1904 when the Society acknowledged practice and explicitly allowed fellows to bring women guests to Ordinary meetings.

One woman who availed herself of such an opportunity, and attended over 50% of the meetings between 1909 and 1912, was Catherine Raisin, Head of Geology at Bedford College from 1890 to 1920 (and from 1891 to 1908 also Head of Botany). A product of North London Collegiate School, where she stayed on as a teacher until she was 20 years old, Catherine always had a fierce interest in women's education. While a student at University College, London, she had helped set up the Somerville Club for women (of which she was Honorary Secretary), which rented rooms in London's Soho and was dedicated to promoting discussion of current issues; it quickly attracted a huge number of members but was disbanded only seven years later, possibly because those driving it had other commitments, rather than the issues having disappeared.

In 1893, for her work in 'petrology and other branches of Geological Science', the Geological Society decided to make an award to Raisin from its Lyell Fund. Charles Lyell's *Principles of Geology* had been her inspiration in girlhood, so, coupled with the fact that she was the first woman to be so honoured, she must have felt a peculiar mixture of pride, sadness, and annoyance when Professor T.G. Bonney had to accept the award on her behalf, women not being admitted to the Society's meetings at that time (Burek 2007).

In 1900, a Lyell Fund prize was again awarded to a woman, this time to Gertrude Elles of Newnham College, 'as an acknowledgement of the value of her contribution to the study of the Graptolites and the rocks in which they occur, and to encourage her in further research'. Elles, like Raisin, was banned from receiving her award in person, despite several attempts by fellows to overturn the rule. Her award was collected by her Cambridge professor, Thomas McKenny Hughes, who commented:

> *I am glad to have been asked to receive the Award from the Lyell Fund for transmission to Miss Elles, who is debarred by circumstances over which she has no control from standing here to receive for herself this mark of recognition which the Council of the Society have bestowed upon her.*[3]

Although Catherine Raisin became a Fellow of the Linnean Society (FLS) in 1906, and all through her years at Bedford College women were allowed attend meetings of the Geological Society and have their papers read for them, it was not until 1919, when she was sixty-four, that she and other women, including Gertrude Elles, were finally admitted as full Fellows of the Geological Society.

Grace Frankland was involved in a scrap with a comparable society; in her case it was The Chemical Society (founded 1841). In what is becoming a familiar story, it was during the 1880s that the question of fellowships for women began to be discussed seriously among the chemists. After his first abortive attempt in 1888 to change the rules to allow women

[3] https://blog.geolsoc.org.uk/.

into the society, William Ramsay, Professor of Chemistry at University College London, who notably employed women research assistants in his own laboratory—although he believed women did their best work when working under the guidance of men—tried again in 1892 (Jones 2009, 91).[4] A woman, possibly Emily Lloyd, was proposed for a fellowship. In spite of support from many sources, including that of his friend Norman Lockyer given through the journal *Nature*, Ramsay's advocacy narrowly failed to win sufficient support (Baldwin 2015, 78). It was the opinion of Henry Armstrong, his main opponent, that

> *...the very women who have shown their ability as chemists should be withdrawn from the temptation to become absorbed in their work for fear of sacrificing their womanhood; they are those who should be regarded as chosen people, as destined to be the mothers of future chemists of ability; it was a view that was shared by a majority of the Council.* (Rayner-Canham and Rayner-Canham 2003)

In 1904 the Chemical Society got itself into difficulties when it admitted Marie Curie as a Foreign Fellow. The proposal to admit her had prompted legal opinion to be sought—not for the first time. The advice as far as British women were concerned was the same as it had been twenty-four years earlier, married women were definitely not admissible, and the Chemical Society's Charter probably also barred single women: Mme Curie (née Sklodowska) could be admitted on the grounds that foreign fellows bore no responsibilities or duties.

Marie Curie's fellowship did, however, encourage women chemists to continue to press their case for admission, and in October 1904 nineteen of them, drawn from various institutions across the country—and led by Ida Smedley (Manchester), Ida Freund (Newnham, Cambridge), and Martha Whitely (Imperial, London), and including Grace Frankland—submitted a petition to the Society:

> *We, the undersigned, representing women engaged in chemical work in this country desire to lay before you an appeal for the admission of women to Fellowship in the Chemical Society....during the last thirty years (1873–1903) the names of about 150 women...have appeared [in the Society's publications]*

[4] Ramsay was later a Nobel Prize winner for his discovery, with Lord Rayleigh, of the 'noble' gases, argon, etc.

> *as authors or joint authors of some 300 papers.* (Cited by Rayner-Canham and Rayner-Canham 2008, 64–65)

The proposal was voted down, not by Council but by those Ordinary members who made the effort to attend an Extraordinary Meeting of the Society to consider the matter. The struggle continued on and off for a further fifteen years and it was only in the post-war world of 1920, forty years after discussions began, that women were finally admitted to the Chemical Society.

Helping the Chemical Society to change its mind, and finally admit women, was parliament's removal in 1919 of the Sex Disqualification Act. Scientific societies, like potential employers of women, were forced by the Act to consider whether or not they were discriminating on grounds of sex or marital status. Nevertheless, with the help of clever lawyers and some sophistry, many found ways around the new legislation.

The Chemical Society (founded 1847) should not be confused with the Royal Institute of Chemistry (RIC: founded 1877), whose focus was much more on qualifications, standards and professional status; membership was dependent upon applicants first passing an examination. The RIC was notoriously tricked into admitting women when in 1892 Emily Lloyd, whose chemical studies had taken her to Aberystwyth, Birmingham, and UCL, entered herself for the RIC's Associateship examination, albeit under the gender-neutral title 'E. Lloyd'. Emily passed the exam, forcing the RIC, somewhat reluctantly, to admit her and other women (Rayner-Canham and Rayner-Canham 2008, 56–57).

When the Biochemical Society (initially 'Club') was formed in 1911 it spent a year discussing membership for women before deciding to admit them, which was still several years before the much older Chemical Society finally changed its rules.

The British Mycological Society (formed 1896), in which two of Linnaeus' Ladies, Gulielma Lister and her good friend Annie Lorrain Smith, were to be prominent members, was open to women from the outset. In the list of 'Foundation Members' published at the front of the first volume of the Society's *Transactions*, two women are listed: Mrs Massee, the wife of the President, and Miss EA Rose, [presumably] the daughter of another member, John Rose, President of the Worcestershire Naturalists Trust (Anon. 1896a). Later in the same volume (p. 86), there are the names of ten further members who have joined 'during the Season

1897–98'; it is a list which includes Annie Lorrain Smith and one other woman, a Mrs L. Montague.

The broad picture presented by a clutch of learned and scientific societies is then that in 1904–1905 the Linnean Society was neither extraordinarily early in admitting women, nor particularly late. The event was important however, with a significance well beyond the Linnean, because of that Society's age and status, and the fact that it had been targeted by Lady Warwick's Association.

While most scientific and learned societies had their base in London, drawing their members mainly from London and its surrounding areas, there were two other types of institution which each had a regional basis and which, in some cases, were welcoming to women: Literary and Philosophical Societies (the 'Lit. & Phils.'; a 'natural philosopher' being an early name for a scientist) and the Field Clubs.

Literary and Philosophical Societies

Most Lit. & Phils. originated in the newly industrialized cities, just like the astronomical societies mentioned earlier. While addressing the desire of the new urban middle classes for self-improvement, their emphasis on science, especially applied science, reflected an increasing closeness of science and industry. Manchester's Lit.& Phil. (founded 1781) listed among its members such scientific 'greats' as James Joule, John Dalton, and Ernest Rutherford; not to be outdone, Newcastle's (founded 1793) early Presidents included Robert Stephenson, William Armstrong, and Joseph Swan. In spite of their positive attitude to scientific progress and the advancement of knowledge, the Lit. & Phils.' attitudes to women tended to be conservative. Thus, Manchester did not admit women to full membership until the twentieth century.

Lamenting the lack of opportunity for direct participation in scientific conversation, in 1866 Lydia Becker set about organising a Manchester Ladies' Literary Society (see the quotation at the opening of this chapter). Somewhat ironically, on the strength of her address to the BAAS in 1868, she was invited to speak at the 'Hull Literary Institute and the Nottingham Philosophical Society' [*sic*] (Gianquitto 2013; Parker 2001).

Although in Newcastle women had technically always been able to join the society as ordinary members, and one had done so by 1801, a new sub-class of 'Reading Members' was established in March 1799, particularly exempting a woman's candidature from the usual procedure of being proposed and voted on. Sheffield's Lit. & Phil. was slightly younger and decided at its very first meeting, in 1823, that its meetings were open to women if they were accompanied by male members of the society (who paid the not insubstantial entrance fee of two guineas, followed by an annual fee) (Purvis 1991, 98). Women were enabled thereby to attend lectures on subjects such as electricity, mechanics, and optics. From 1869 the rules were relaxed further and lectures were occasionally given by women, for example Mary Kingsley of the subject of her travels in West Africa. When Emily Davies embarked on a campaign to raise funds for the establishment of Girton College, Cambridge, she included in her tour the Nottingham's Lit. & Phil. Society.

As in the case of Leicester (founded 1835), a Lit. & Phil. might have a library and also a museum for housing collections, such as of fossils, where from 1870 professional lecturers, such as Thomas Henry Huxley and other fellows of the Royal Society, might be invited to speak. Leicester, however, did not admit women as full members until 1885, an all too familiar story.

In *The Naturalist in Britain*, David Allen argues that in Victorian times it was the Lit. & Phils. that gave rise to a new type of institution, the Field, or Natural History, Club, or Society. The most obvious difference was that the emphasis of these newer groups was on outdoor activities, typically involving whole-day excursions rounded off with a convivial meal and drinks (Allen 1976, 143–145).

Field and Natural History Clubs

Botany and, to a lesser extent, geology were popular pastimes for women in the Victorian age, satisfying that urge felt by both sexes to collect and classify objects from the natural world. Botanising in particular was widely approved by polite society as being a suitably genteel and healthy occupation for the fairer sex—as was so well documented by Anne Shteir in her book *Cultivating Women, Cultivating Science* (1996). Plants were not as inescapably sexualised as were animals. Following exercise in fresh air, when plants were gathered, women could classify their collections, perhaps use a microscope to study their specimens, exchange rarities with friends, and even establish their own herbaria. Many chose to draw and paint the flowers they had collected. When the Botanical Society of London began life in 1836 women formed 10% of its founder members.

Industrialisation, with the accompanying mass transfer of population from the country to the growing towns and cities, merely served to promote in both sexes an interest in the natural world which, in turn, generated a plethora of 'field clubs', again deriving strength from their being organised on a local basis. Some were popular to an extent that today is barely imaginable: one excursion of the Manchester Society attracted 550 people (Allen 1976, 148). Equally difficult to imagine is that, given such crowds, much serious study was done. Where women were admitted their presence was all too often resented, or they were admitted as 'visitors' on only a few selected days in the year.[5] The Rev. Charles Kingsley, who founded the Chester Natural History Society, famously remarked:

> *Those good ladies quite spoilt my day—but what can you do? When they get to certain age you must either treat them like duchesses or shoot them.* (Cited by Allen 1976, 151)

Not all such organisations were open to women, an example of the latter being the much smaller Woolhope (Naturalists Field) Club, which began life in the city of Hereford in 1851 but did not admit women until 1919. The Woolhope is important because it was from that body that the British Mycological Society (BMS) can trace its origins. The Woolhope was devoted to natural history and geology, as well as archaeology. It placed a strong emphasis on excursions into the countryside of Herefordshire and virtually invented 'fungus forays', attracting the best mycologists from around Britain to its meetings. It was only when interest in mycology waned within the Club that, in the early 1890s, leading mycologists transferred their allegiance to the Yorkshire Naturalists Union (YNU), before subsequently peeling off from that organisation in 1896 to form their own, specialist, BMS. Among those who helped found the BMS was Worthington G. Smith FLS, who was also a founder and exceptionally active member of the Essex Field Club.

The 'Epping Forest and County of Essex Naturalists' Field Club' or, as it was more conveniently called after 1882, 'The Essex Field Club', was dedicated to a study of the botany and zoology, geology, and

[5] Average attendance at monthly meetings of The Winchester Natural History Society in 1872 was 26, plus 'visitors', mainly females. The Society had been founded two years earlier by the Rev. CA Johns at the suggestion of his friend, the Rev. Charles Kingsley (Dare and Hardie 2008, 156).

anthropology of the county of Essex, with special emphasis on Epping Forest. Its life, like the Woolhope's, was centred on excursions, or as The Essex called them, 'field meetings' (Fig. 2.1). Their purpose was twofold:

> *(1) to promote friendliness and interchange of ideas among the members under pleasant conditions of out-of-door rambles and excursions, and*
> *(2) to afford opportunities for scientific demonstrations in the field, by skilled Conductors, of subjects of interest to biologists, geologists, and antiquaries.* (Anon. 1906, 2)

At the inaugural meeting in 1880 an annual membership fee of fifteen shillings for men and ten shillings for ladies was proposed (with suitable reductions for members living more than fifteen miles radius from the Club's headquarters). The first field meeting was held at Ongar on 29 March 1880 and for 'a very substantial lunch and tea' members were

Fig. 2.1 Over 40 men and women of the Essex Field Club gathered at Tyler's Common, near Upminster, on 26 July 1890. (Source: Permission of the Essex Field Club)

charged a further five shillings, whatever their sex. Meetings were well attended. On Saturday, 3 July 1880, nearly fifty 'enthusiasts' joined an excursion to the 'Ancient Earthworks of Ambresbury Banks, and Loughton'. The day was exceptionally wet forcing the eight lady members and friends to don 'waterproofs' and carry umbrellas. Professor George Boulger FLS FGS and Mr Henry Walker FGS taught the bedraggled party the finer points of the biology, geology, and palaeontology of the areas they were exploring, before everyone retired for a cheering, and warming, 'high tea' at the Forest Hotel, Chingford (Anon. 1881).

Founder members of the Essex Field Club included John Lubbock, Frank Crisp, and Arthur Lister, men who would prove among the keenest supporters of women's admission to the Linnean Society (Anon. 1882). One of the first lecturers was a young entomologist, Mr EB Poulton of Oxford University, who was to be a mainstay of the Essex, and keen supporter of similar clubs (see below), over the next three decades. His talk, 'The Protective Value of Colour and Attitude in Caterpillars', was illustrated by 'a long series of lantern slides' (Anon. 1884a).

Here indeed was a meeting place where women could mix freely with men, each sex enjoying the other's company and interests. Thus, the archives of the Club record that Marian Ridley (the future Mrs Farquharson; she joined in 1881) was in 1884 at the same field meeting as EM Holmes and, also, Professor George Simonds Boulger of the Royal Agricultural College, Cirencester, who was exhibiting specimens on behalf of another leading light of the Linnean Society, John Gilbert Baker, Keeper of the Herbarium at the Royal Botanic Gardens, Kew (Anon. 1884b).[6] Present at a meeting in 1894 was the same Professor Boulger, this time along with Arthur Lister, Miss Gulielma Lister (his daughter and future FLS), John Lubbock, and a Miss Gibbs (presumably Lilian Suzette; another future FLS) (Anon. 1894). On other occasions E.M. Holmes mixed with the Listers, and with Frederick J. Hanbury (Le Lievre 1980, 161).[7] D.H. Scott, Botanical Secretary of the Linnean Society, was one of the 'distinguished botanists' who joined the annual fungal foray of 1905. Sometimes trips were arranged to gardens or houses of interest. The Countess of Warwick frequently threw open her celebrated gardens at Easton Lodge, near Dunmow, to

[6] The Club's publications record only the names of those leading meetings, plus any distinguished guests, not the names of everyone who attended.

[7] Le Lievre states that Gulielma Lister was a cousin of the Hanburys, but the present author can find no evidence for this in Locke 1916.

parties from the Field Club in the 1890s, and in later years annual visits were made to Warley Place, the nearby gardens and home of Ellen Willmott FLS, who joined the Club in 1907 (Le Lievre 1980, 209).

In 1906, the British Mycological Society, led by its current President, Arthur Lister, and itself recently evolved out of a series of field clubs, held its annual foray in Epping Forest, an event organised jointly with the Essex Field Club (Lister 1906).[8]

Connections between the Club and the history of the Linnean Society are underlined by the genesis of the two museums for which the Club was responsible (Anon. 1908, 8–9). Its first, The Epping Forest Museum, was opened in 1895. Housed in part of Queen Elizabeth's Lodge at Chingford, a picturesque Tudor building, it displayed objects of interest concerning the history, geology, and natural history of the forest. The free museum was so popular with the public—among its first visitors being a party from the Toynbee Hall Natural History Society[9]—that in 1899 the Epping Forest Committee of the Corporation agreed to renovate the building at the cost of £1200 so that all of it could be given over to the museum (Anon. 1896b, 21). The architect chosen to take charge of the reconstruction was John Oldrid Scott, brother of D.H. Scott.

The second and larger museum, The Passmore Edwards (named after its major benefactor), was sited in new premises at Stratford. It incorporated the old Chelmsford Museum, housed the Club's main reference collections, and from 1900 became its headquarters. At the formal opening, on the 18th of October of that year, a bust of Passmore Edwards was unveiled by the Countess of Warwick, who had been a member of the Essex Field Club since 1896.

Another body which brought together several Lady-Fellows was the Holmesdale Natural History Club, which held its meetings in Reigate, Surrey, and occasionally in nearby Redhill. Ethel Sargant appears to have been the key figure here for, as will be seen, she not only served as an officer (Treasurer) and Council member of the Club but her name, like those of Rina Scott and Annie Lorrain Smith, appears on various programmes of lectures, as too does that of Winifred Smith, who was elected FLS shortly after the period considered in this book.

[8] At the same meeting, Annie Lorrain Smith was elected to be the next President, and Helen Fraser and Evelyn Welsford (Chap. 5) were admitted as new members.

[9] Toynbee Hall was named in memory of Arnold, the brother of Grace Frankland (née Toynbee, see Chap. 7).

The Club's *Proceedings* record that on 4th December 1903

> *Miss Ethel Sargant then delivered an interesting lecture entitled, 'The Effect of the Seasons on Plant Life tracing the modifications of habit undergone by seedlings of various plants to enable them to cope more effectually with adverse climatic conditions'. The lecture was illustrated with diagrams and with actual specimens.*

As recorded in the *Proceedings*, the Club's meetings were however sometimes more light-hearted:

> *A soiree was held in the Small Hall Redhill on April 24 1903, when about 42 members and friends spent a most enjoyable evening (songs, violin solos, lantern views of Surrey scenery and a display of 'Electrical Fireworks', plus numerous exhibits). The Hall was tastefully decorated with plants kindly lent by Miss Ethel Sargant and Miss M C Taylor.*

Natural history clubs abounded in the nineteenth century, catering for their members' recreation and, to varying degrees, their education. Clubs at that end of the spectrum where emphasis was on learning might have libraries, museums, and reference collections, and, as in the case of The Essex, publish a journal and occasional studies (which it named 'Special Memoirs') featuring their own region or discipline.[10] They attracted men and women who were serious about their science, including many from that new breed of laboratory-based professionals who were anxious not to lose touch with the natural world.

British Association for the Advancement of Science (BAAS)

There is one further place of importance where a woman could meet like-minded men, and women too. It was at the annual meeting of the BAAS. From its very first meeting in York, in 1831, women were free in principle to attend the BAAS' meetings (Higgitt and Withers 2008). In practice they were guests, for during those first few years women were welcome only at the side-events, conversaziones, and dinners, which formed a part of the week-long meetings, held each year in a different city. Women were not

[10] For example, Meldola and White (1885) reported on the previous year's earthquake in East Anglia.

welcome at the Section Meetings where research papers were read. However, by 1837 they were able to attend meetings of the Geology Section (C), when the President, Adam Sedgwick, reported that over 300 people filled the galleries each day. In the same year, women were also allowed to attend the botanical sections of the Biology Section (D), the zoological parts being deemed inappropriate. Happily, in 1839, the rules were changed so that women could attend all sectional meetings, and in 1848 they were admitted as members.

As part of the 1878 meeting in Dublin a geology excursion was organised for women, something which, as seen in the next chapter, was a novelty for the times. *The Irish Daily News* gleefully contrasted the unfeminine appearance of the women who wielded geological hammers with the appearance of those women who 'looked on, strove to look learned, and sighed' (Higgitt and Withers 2008).

Whatever the weaknesses of these week-long BAAS meetings, and they were described by one cynic as 'a philosophers' picnic', they had by the middle decades of nineteenth century helped nurture many women's interest in the natural sciences, an interest which later in the same century was to evolve into a more active participation. The popularity of BAAS meetings increased steadily for both sexes, so that the Manchester meeting of 1887 was attended wholly or in part by some 3838 people. Many were schoolteachers but the great majority were members of the general public seeking entertainment along with self-improvement, Section E (Geography) often attracting the largest crowds, particularly to talks given by renowned African or polar explorers, such as David Livingstone or John Franklin.

Like schoolteachers, the clergy were always well represented at BAAS meetings for the 'Parson-Naturalist' was a well-recognised character in rural life, several of them delivering papers based on an unique knowledge of their local flora, fauna, and geology which they had accumulated over many years (Armstrong 2000, 176). The Reverend Charles Kingsley was one such man, another was the Reverend Francis Orpen Morris, vicar of Nunburnholme in the East Riding of Yorkshire, author of many popular books on ornithology, including *Natural History of the Nests and Eggs of British Birds*, 1853, which proudly announced in its frontispiece that he was a member of the Ashmolean [Natural History] Society of Oxford. Morris's nephew, and likely pupil, was Frederick Orpen Bower, who as a young man was a friend and contemporary of Dukinfield Henry Scott at the Jodrell Laboratory before, in later life, being made Professor of Botany at the University of Glasgow. Morris' church was restored in 1871 by the architect George Gilbert Scott, Jr, older brother of DH Scott (Morris 1897, 53). BAAS meetings were mixing places without equal.

If professional scientists formed only a small part of the total attendance, it was they who manned all the committees [literally!], led discussions, and delivered papers. It was not until 1860 that the first paper was read by a woman, Mary Carpenter, which was to the Sociology Section. In 1868, Lydia Becker, the botanist and leading suffrage campaigner, delivered to the Biology Section (D) of the BAAS a paper, 'On the alteration in the structure of Lychnis diurna (Red Campion) observed in connection with a parasitic fungus' (Parker 2001). Becker showed that infection changed female into hermaphroditic flowers. It represented for her clear evidence for plasticity in the development of organisms, something that was fundamental to her belief that the human mind has no sex; differences, she argued, were the result of differences in education and cultural expectations (Gianquitto 2013). Becker's research was controversial in another way for, in dealing with sexual matters, it challenged many men's, and some women's, views of what was an acceptable area of study for a woman.

Progress towards equality of the sexes was slow but a major step was taken at the Bradford meeting in 1900 when, 'It was agreed women are eligible for General and Sectional Committees' (Anon. 1901). It may well have been Hertha Ayrton's lecture 'L'intensité lumineuse de l'arc à courants continus' at the International Electrical Congress in Paris in the same year that helped Marcus Hartog, her cousin, persuade the British Association to allow women onto their committees (Mason 2004).

BAAS programmes were at last featuring more women. Thus, in 1902, papers were given in the Botany Section (K) by Margaret Benson, Rina Scott, and Annie Lorrain Smith, all among the first female fellows of the Linnean; in 1903, the names of Ethel Sargant and Edith Saunders, Agnes Robertson, Marie Stopes, and Ethel Nancy Miles Thomas appear, as in 1904 do those of Rina Scott, Marie Stopes, and Helen Fraser (all of whom were at some time made FLS). The prestige of the BAAS was never higher than in 1904 when, at its Cambridge meeting, the Presidential Address was given by the Prime Minister, Arthur J Balfour.[11]

[11] Among those listening to Balfour was Clara Collet, a statistician and social scientist.She was not out of place since the BAAS, which she had joined in 1890, had had since 1856 a Section (F) for Social Science and Statistics. Educated at North London Collegiate School,

BAAS meetings not only encouraged potential natural scientists to take their interests more seriously, they provided also an avenue whereby the slowly growing number of committed females could meet with, and become known by, their male counterparts. Not least, in the gentlest of ways, it accustomed those men to the presence of women in a scientific setting—a case of familiarity breeding respect.

Personal Connections

It was a relatively small group of *men* who were the leaders of the world of natural history in the last decades of the nineteenth century and it was common for them to belong to several societies or groups. Thus, for example, inspection of the online records of the Essex Field Club for this period reveals 'FLS' was appended to the names of many of those organising or attending meetings. When the mycological activities of the YNU were at their zenith in 1895, both the President and the Secretary were fellows of the Linnean.

There was no absolute distinction between scientific and learned societies, on the one hand, and field clubs, on the other. The former tended to be more formal and with grander ambitions, sometimes organising and financing international expeditions. Field clubs typically had a more local focus, although the better funded among them set up their own museums and reference collections, as in the case of the Essex Field Club. Both types of organisation often involved the same group of male officers and leaders, men who also organised and presented papers at meetings of the BAAS. Furthermore, there were direct connections between the field clubs, facilitated by the 'Corresponding Societies Committee' of the BAAS. Each year representatives of the clubs and societies, such as the Essex and the Yorkshire Naturalists, would get together at the annual meeting of the BAAS to discuss cooperation on projects of mutual interest. At the Bradford meeting in 1900, the Club's delegate and Chair of the Committee was Professor EB Poulton FRS FLS, someone long connected with the Essex Field Club (Poulton 1899–1900). By 1904, in Cambridge, over thirty local societies were represented; Ethel Sargant, representing the Holmesdale Society, mixed with, amongst others, the Rev. TRR

Clara became a Fellow of the Royal Statistical Society in 1892, and in 1896 was made UCL's first female Fellow (McDonald 2004, 194–195).

Stebbing FLS, representing Tunbridge Wells, and George Claridge Druce FLS, representing the Ashmolean Society of Oxford (Chap. 6).[12]

For women wishing to join the mainstream of scientific society, such overlap and inter-connectedness of local groups, which BAAS meetings encouraged, meant that when they were allowed to participate in local activities, either as full members or merely as guests, there was a good chance that they could become acquainted with one or more male fellows of the national societies they targeted. If a positive impression was left with a fellow, there was a good chance that he would tell his friends in other groups about that women. As a result, the number of men who might accept the general principle of opening their societies to women gradually swelled. And the path towards a fellowship for a particular women was thereby smoothed.

This chapter should not close with a completely rosy picture, for a *caveat* must be added. The capacity of any woman to interact with others at a meeting of a society or field club, and to contribute to or influence its business, was affected by her level of education, particularly in her chosen science. The harsh reality was that, however enthusiastic, self-confident, and sociable she might be, her interactions, and whether or not she was accepted by the men she viewed as her equals, would have been limited by the extent of her knowledge, and familiarity with the vocabulary, of the science she favoured. Women's educational experiences, especially in the sciences, were critical to their acceptance into scientific society. They are the subject of the next chapter.

References

Allen, D.E. 1976. *The Naturalist in Britain. A Social History*. Princeton: University Press.

Anonymous. 1881. *Essex Field Club Journal* 1: 24–27.

———. 1882. *Essex Field Club Journal* 2, Appendix: xx.

———. 1884a. *Essex Field Club Journal* 4: clxxxvi.

———. 1884b. *Essex Field Club Journal* 4: 194.

———. 1894. *Essex Naturalist* 8: 53.

———. 1896a. *Transactions of the British Mycological Society* 1: 3–5.

———. 1896b. *Essex Naturalist* 10: 21.

———. 1899. *Nature*, October 26.

[12] Conference of Delegates, BAAS 136, Special Collections, Bodleian Library, University of Oxford.

———. 1901. *British Association for the Advancement of Science, Annual Report*, cvi.
———. 1906. The Essex Field Club: What It Is: What It Has Done 1880–1906; and What It Wishes to Do. *Special Publication of the Essex Field Club*, 2.
———. 1908. *Essex Field Club: Yearbook and Calendar for 1908.*
Armstrong, P. 2000. *The English Parson-Naturalist. A Companionship Between Science and Religion*. Leominster: Gracewing.
Bailey, Mandy. 2016. Women and the RAS: 100 Years of Fellowship. *Astronomy and Geophysics* 57: 19–21.
Baldwin, Melinda. 2015. *Making "Nature". The History of a Scientific Journal.* Chicago: University Press.
Burek, Cynthia B. 2007. The Role of Women in Geological Higher Education – Bedford College, London (Catherine Raisin) and Newnham College, Cambridge, UK. In *The Role of Women in the History of Geology*, ed. Cynthia B. Burek and Betty Higgs, 9–38. Geological Society, Special Publications, 281.
Chapman, A. 2016. The Lady Astronomers of Victorian Britain. *Astronomy and Geophysics* 57: 12–13.
Dare, Deirdre, and Melissa Hardie. 2008. *A Passion for Nature. 19th Century Naturalism in the Circle of Charles Alexander Johns.* Penzance: Patten Press.
Desmond, A. 1994. *Huxley: The Devil's Disciple.* London: Michael Joseph.
Gianquitto, Tina. 2013. Botanical Smuts and Hermaphrodites; Lydia Becker, Darwin's Botany and Education Reform. *Isis* 104: 250–277.
Goodman, M. 2007. *Suffer and Survive. The Extreme Life of Dr. JS Haldane.* London: Simon & Schuster.
Herries Davies, G.L. 2007. *Whatever Is Under the Earth. The Geological Society of London 1807 to 2007.* London: The Geographical Society of London.
Higgitt, Rebekah, and C.W.J. Withers. 2008. Science and Sociability. Women as Audience at Meetings of the British Association for the Advancement of Science, 1831–1901. *Isis* 99: 1–27.
Jones, Claire G. 2009. Bodies of Controversy? Women and the Royal Society. In *Femininity, Mathematics and Science, 1880–1914*, ed. Claire G. Jones, 175–204. Berlin: Springer.
Le Lievre, Audrey. 1980. *Miss Willmott of Warley Place. Her Life and Her Gardens.* London: Faber & Faber.
Lister, A. 1906. The Epping Forest Foray: 1st to the 6th October 1906. *Transactions of the British Mycological Society* 2: 133–141.
Locke, Audrey. 1916. *The Hanbury Family*, 2 Vols. London: Arthur L Humphreys.
Mason, Joan. 1991. Hertha Ayrton (1854–1923) and the Admission of Women to the Royal Society of London. *Notes and Records of the Royal Society of London* 45: 201–220.
———. 1995. The Women Fellows' Jubilee. *Notes and Records of the Royal Society* 49: 125–140.

———. 2004. *Ayrton [née Marks] Sarah [Phoebe] Hertha (1854–1923)*. Oxford Dictionary of National Biography Online. https://doi.org/10.1093/ref:odnb/37136.

McDonald, Deborah. 2004. *Clara Collet 1860–1948. An Educated Working Woman*. London: Woburn Press.

Meadows, A.J. 2008. *Science and Controversy. A Biography of Sir Norman Lockyer, Founder of Nature*. London: Macmillan.

Meldola, R., and W. White. 1885. Report on the East Anglian Earthquake of April 22nd 1884. *Essex Field Club Special Memoirs*, 1. London: Macmillan.

Morris, M.C.F. 1897. *Francis Orpen Morris. A Memoir*. London: John Nimmo.

Parker, Joan E. 2001. Lydia Becker's 'School for Science'. *Women's History Review* 10: 629–650.

Poulton, E.B. 1899–1900. The Corresponding Societies Committee of the British Association, Bradford, 1900. *Essex Naturalist* 11: 310–311.

Purvis, June. 1991. *A History of Women's Education in England*. Milton Keynes: Open University Press.

Rayner-Canham, Marelene F., and G.W. Rayner-Canham. 2003. Pounding on the Doors: The Fight for Acceptance of British Women Chemists. *Bulletin for the History of Chemistry* 28: 110–119.

———. 2008. *Chemistry Was Their Life. Pioneering British Women Chemists, 1880–1949*. London: Imperial College Press.

Tansey, Tilli. 2015. Women and the Early Journal of Physiology. *Journal of Physiology* 593: 347–350.

CHAPTER 3

Educational Opportunities for Girls and Women

Speaking broadly, a man ought to know any language or science he learns, thoroughly—while a woman ought to know the same language or science, only so far as may enable her to sympathise in her husband's pleasures, and in those of his best friends.
—*John Ruskin (1865, 153)*

Until the mid-nineteenth century, the few girls who received any sort of education, whether it was serious or perfunctory, usually did so within the confines of their family home. Until well into the 1870s and 1880s, a significant proportion of even those women who did successfully complete a university course had received all their prior education privately (Dyhouse 1981, 45). They were taught directly, or sometimes guided to teach themselves, by parents, tutors, or governesses who had little or no knowledge of the sciences. In girlhood, they were isolated and alone, unless they were put together with a sister or brother, probably of a very different age. If family finances were limited, it was upon the education of sons, not daughters, that money was spent, though brothers on holiday from boarding school might pass on their newly acquired knowledge to their stay-at-home sisters (Dyhouse 1981, 15). An unique insight is provided by Molly

P. Ayres, *Women and the Natural Sciences in Edwardian Britain*, Palgrave Studies in the History of Science and Technology, https://doi.org/10.1007/978-3-030-46600-8_3

Hughes', *A London Girl of the 1880s*; she did not enter North London Collegiate School until she was 16, in 1883, and tells charmingly how a better educated brother helped fill the gaps in her rudimentary knowledge of algebra and geometry (Hughes 1978a, 62).

Girls who were lucky enough to be sent to a school or college were often ill-prepared by their background for the education from which they might benefit (Raikes 1908, 35). A Cambridge tutor of the future astronomer, Annie Maunder (née Russell), remarked that in spite of her having attended The Ladies' Collegiate School in Belfast, she was 'more than ordinarily handicapped—even for a woman—by an insufficiency of preliminary training' (Bailey 2000). Annie nevertheless finally triumphed for she was to win a paid position at the Greenwich Royal Observatory, something quite exceptional in the 1890s.

Such limited educational experiences meant that those few women who somehow developed an interest in the natural sciences were all too often restricted to being collectors of plants, insects, fossils, or rocks, or artists who drew the same. They were effectively isolated, save for any correspondence they might have with more knowledgeable male experts.[1] It was necessary that social attitudes to the education of females should change before they could enjoy the more specialised instruction which would equip them to become actively involved in the sciences. And attitudes, such as John Ruskin's, would not change, and girls' horizons would remain limited to pleasing their future husbands, so long as the age-old pattern of home education prevailed.

Signs of change were apparent when the very first volume of the journal *Nature* appeared in 1869. In a brief article entitled 'Lectures for Ladies', its editor, Normal Lockyer, noted that such lectures were springing up in cities across Britain, often under the auspices of Ladies' Educational Associations (LEA) (Lockyer 1869). *Nature* and Lockyer followed up the next year, 1870, in an article 'The Scientific Education of Women' by deploring how the natural sciences were under-represented in such lectures, albeit with a few notable exceptions, such as a programme of lectures by professors Huxley, Guthrie, and Daniel Oliver to women at the South Kensington Museum on physiography (physical and mathematical geography), physics, and botany (Lockyer 1870).

[1] Gianquitto 2013, 255, notes that many men were happy to give advice and discuss ideas. Thus, Lydia Becker included Charles Darwin, Alfred Russell Wallace, and Charles Babington (Professor Botany, University of Cambridge) among her correspondents.

It was in the 1870s and 1880s that significant changes occurred as the number of schools, colleges, and universities that were open to girls and women slowly increased. In spite of widespread prejudice (as in the quotation above), increasing numbers of young women were introduced to the fundamentals of the natural sciences. With such knowledge, and armed with a working vocabulary, they could at last become active participants in those sciences, hoping that they might eventually contribute to the fund of human knowledge about the natural world, and interact on an equal basis with their male peers. This chapter traces the critical development of the schools, colleges, universities, and laboratories which provided new openings for those women who at the turn of the century were knocking on the doors of scientific societies.

Schools and Colleges

London's Harley Street was in 1848 already building a reputation for being a magnet for the most prestigious consultants in medicine and surgery. In that year, however, there opened at number 66 (later renumbered 45) a very different and ground breaking establishment called Queen's College. Originally aimed at the education of future governesses, its remit was very soon broadened so that it was, 'for the instruction of ladies generally' (Anon. 1969, 311–312).[2] Open to girls and women above the age of twelve, it sought principally to train the teachers without whom any expansion of women's education would always be constrained. Day-time courses attracted fees, whereas evening classes, which were particularly suitable for working governesses, were free. In addition to qualifications in the traditional arts subjects, certificates were offered in geography, geology, natural philosophy [natural sciences, especially the physical sciences], and botany, though there is no record of whether these last options were taken up (Raikes 1908, 28). The professors of King's College, London, contributed greatly to teaching and administration during the early years of Queen's.

Prevailing standards of propriety required that girls should be chaperoned during classes because their teachers were male, a task undertaken by 'Lady Visitors', among whom was Lady Henrietta Maria Stanley, one of the founders of Girton College, Cambridge, along with Emily Davies and

[2] See *British History Online* http://www.british-history.ac.uk/vch/middx/vol1/pp311-312.

Barbara Bodichon. At Bedford College, London (see below), where students were in the main a little older, a system of Lady Visitors was similarly in operation. Molly Hughes, who in adult life helped establish at Bedford a department for training teachers, remembered, 'Sharp criticism by a young professor…would often unnerve a girl and reduce her to tears; so the Lady Visitor was present [also] to prevent such a break-down, or to calm the student with smelling salts' (Hughes 1978b, 16). One of the greatest tasks faced by Hughes and her colleagues was instilling in their students the self-confidence they would need if they were to survive in a classroom.

The chairman of the governors and inspiration behind Queen's was the Rev. Frederick Denison ('FD') Maurice, Professor of Theology at King's and founder of the Christian Socialist movement. He had tested his ideas in 1847 by delivering one of the earliest series of evening 'Lectures for Ladies'. Shy and self-effacing, FD Maurice was nevertheless an intellectual leader among his contemporaries, one of his more dangerous ideas being the proposal to teach mathematics to women. The Rev. Charles Kingsley supported Maurice in setting up not only Queen's but also the Working Men's College of which Maurice was the Principal from 1854 to 1872 (a Working Women's College followed in 1864).[3]

Queen's is notable for having spawned in its early years such pioneering professional women as Sophia Jex-Blake, who enrolled in 1858 despite her parents' objections, and who was later the first practising female doctor in Scotland. She was in later life a great reformer of medical education, to the benefit of women. In the present context, however, it was three future educationists who hold the greatest interest, Dorothea Beale, later principal of Cheltenham Ladies' College, Frances Mary Buss, founder of the North London Collegiate School for Girls, and Emily Davies, a co-founder of Girton College.

Miss Buss attended evening classes for six nights a week after a day's work teaching at her mother's school. Her immediate object was not merely to learn but to accumulate from Queen's the proof of that learning; proof in the form of certificates that would enable her to charge higher fees in the school which she planned, and thereby make it a successful business. Firm, charismatic, and generating a fierce loyalty among her pupils, Miss Buss wanted her girls to enjoy an education equal to that of

[3] Lord Avebury (John Lubbock) was Principal of the Working Men's College from 1883 to 1898.

their brothers. Inspired by what she had experienced at Queen's, determined to include science in any curriculum, and with an energy that belied her tiny stature, Miss Buss proceeded to start her own school in the family home in Camden, north London. It involved both her father, Robert, and brother, Septimus, as teachers (Rappaport 2001, 117–119). Until then her father had been an artist whose claim to fame was that he had helped illustrate Dickens' novels, but in his daughter's school he was soon teaching science, his lessons including a well-remembered series of chemistry experiments enjoyed by his admiring pupils for their smells and explosions (Rayner-Canham and Rayner-Canham 2008, 16). As her school expanded, reaching 500 pupils by the mid-1880s, it moved into a succession of larger premises. Science was very firmly in Miss Buss' curriculum, though, as in the case of human biology taught under the guise of 'the Laws of Health', it was often made more palatable to the girls by being taught in an applied rather than a pure context (Hughes 1978a, 43).

Miss Buss, along with Miss Davies, was at the centre of two major steps forward in the education of girls. Thanks to their campaigning, a trial was conducted in 1863 at Miss Buss' North London Collegiate School in which girls sat the Cambridge Local Examinations—which were normally open only to boys. On the strength of the girls' excellent results, the Cambridge Local Examinations Syndicate was finally persuaded by the two women to open their examinations from 1868 onwards to girls across the country. Henceforward, girls would have the opportunity to prove they were just as good as their brothers. Whereas these 'Junior Locals' were taken, normally, by sixteen-year olds, the so-called 'Highers' were introduced in 1869, initially for women over eighteen who wished to become teachers.

Other important developments were afoot in 1868, for the Schools Enquiry Commission, the 'Taunton Commission', investigating opportunities for secondary education, found that those for girls were very limited. (It was due to Emily Davies' pressure that girls were included in the enquiry.) Thanks to the Taunton Commission, new endowed schools for girls opened, drawing upon, often ancient, endowments which had been intended for the education of both boys and girls but which had seldom been used for girls. The Commission's recommendation that every large city should have a secondary school specifically for girls resulted in the formation four years later of a new charitable organisation, the Girls' Public Day Schools Company (1872). A dozen such schools were successfully launched in the 1870s in cities like Norwich and Nottingham, all under the auspices of the Company, and modelled largely on Miss Buss' North London Collegiate. Their fees were modest and they included science in their timetables.[4] It was from such schools that the future chemists Ida Smedley (King Edward VI School, Birmingham) and Martha Whitely (Kensington High) emerged, as did a number of other women who feature in these pages.

Both Miss Buss and Miss Beale grew up in large families but, whereas the former was the daughter of an impecunious artist and had to leave school at fourteen, the latter's family was financially more comfortable and Beale was educated mainly by a long line of governesses. At the age of sixteen, in 1847, she was sent to a finishing school in Paris, which she claimed caused atrophy of her thinking powers. Escaping the chaos of the French Revolution in 1848, she quickly enrolled for courses at the new Queen's College. She was such an outstanding pupil, with an aptitude for mathematics, that at the age of 18 she was invited to become the first female member of the teaching staff. If their teacher's youth was in itself not sufficient inspiration for her pupils, it would have been a revelation to them that a woman could teach mathematics. As the years went by minor irritations accumulated and Beale resigned her post in 1856 citing, among a list of reasons, that too much teaching was done by men and too little by women, 'though some classes may be profitably undertaken by men, the education of girls as a whole must be in the hands of their own sex' (Anon. 1969, 311–312). After a brief spell as Head of the Clergy Daughters' School at Casterton, North Yorkshire, she was in 1858, aged only twenty-seven, appointed Principal of Cheltenham Ladies College (Rappaport 2001, 55–56; Buxton 2008, 8).

Cheltenham was for daughters of the upper social class. Miss Beale had no hesitation in telling the Schools Enquiry Commissioners that she would not admit a girl from the 'lower class of society' (Buxton 2008, 8).

In her paper entitled '*On The Education of Girls*', read at the Social Science Congress of 1865, Dorothea Beale outlined what she saw as the critical flaw in the upbringing of young girls: 'I think that the education of girls has too often been made showy, rather than real and useful… thus, while temporary pleasure and profit have been sought, the great moral ends of education have too often been lost sight of'. She was a deeply religious woman.

[4] Emily Berridge and Vida Latham, see Chaps. 5 and 9, respectively, studied at such schools.

Unlike her admissions policy, her educational policy did not please all her school governors, although gradually they were won around to the principal of science and mathematics taking their place alongside the traditional 'accomplishments' of art, music, and needlework. The first chemistry laboratory was opened in 1875. Geography was used as a subtle way of introducing applied mathematics and scientific concepts into the curriculum. A shortage of trained teachers meant that in the early years of Miss Beale's rule many of her staff had to 'double-up', for example teaching both chemistry and Latin. When Mary Russell (the future Duchess of Bedford) entered Cheltenham in 1876 she found that outside lecturers were employed to teach mechanics and hydrostatics (Buxton 2008, 16). Recognising that the growing number of new schools for girls, such as those set up through the Day Schools Company, would flourish only if there was an adequate supply of skilled, trained teachers, Miss Beale introduced in 1876 formal training for secondary teachers, preparing them for the Cambridge Teachers' Examinations, which was probably one of her most important innovations (Green 2013, 34).[5] Staff were now being appointed specifically to teach the sciences, including botany, chemistry, physiology, and zoology. The growing reputation of the school meant that Miss Beale could soon attract able and experienced teachers, women who would inspire their pupils; women such as Charlotte Louise Laurie who, appointed in 1880, wrote *Flowering Plants. Their Structure and Habitat* (1903, Allman & Son), a book beautifully illustrated by her colleague, Winifred Lily Boys-Smith. The two women also collaborated to publish *A Text-book of Elementary Botany* (1905) and *Field Botany* (1906), the latter, which was interleaved with blank sheets for notes made in the field, strongly suggesting that Laurie and Boys-Smith's pupils were taken into the countryside beyond the confines of Cheltenham College. Through the example of their achievements, the influence of such schoolteachers cannot be underestimated.

[5] Not only secondary school teachers were needed. WE Forster's Education Act, 1870, provided the first framework for free, compulsory education for all children up to age 13. Applying to both boys and girls, its impact was mainly on children from working class homes, although it required a supply of well-educated teachers, who came typically from middle class backgrounds.

Higher Education

In 1849, just one year after Queen's College was founded, another event of great significance occurred in London, the opening of the 'Ladies' College' in Bedford Square, Bloomsbury, part of the Duke of Bedford's estates. It did not immediately attract the number of applicants expected by its wealthy sponsor, the social reformer and staunch Unitarian, Elisabeth Jesser Reid. Bedford was for many years beset with financial worries. However, thanks to input from several staff from University College, London (UCL), such as Augustus De Morgan (Professor of Mathematics), who idealistically gave their time to teach the older students without receiving payment, there was created in Bedford Square the first institution in the United Kingdom to offer higher education exclusively to women.[6]

In 1877 Bedford College introduced entrance examinations, helping to define itself thereafter as a place of higher education, rather than a school. While the 'junior department' was allowed to die, Bedford College emerged and eventually prospered.[7] When, in 1878, London University admitted women to its degree courses, Bedford's students could win full bachelor's degrees, rather than mere certificates as previously.[8] Between 1876 and 1885, professorships in chemistry and physics, physiology, and botany and geology were instituted and a new laboratory block, the Shaen Wing, was opened in 1891. The teaching of botany, geology, and chemistry were though heavily intertwined. It was not until 1907 that the then head of the geology department, Catherine Raisin, decided that being simultaneously the head of the botany department was too much for one woman.[9] A recently appointed lecturer, Ethel Thomas (of whom there will be much more later), was the beneficiary, being promoted in 1908 to replace Raisin as head of botany (Burek 2007). In the same year the rapidly expanding college leased eight acres of Crown Land in Regent's Park

[6] The mathematician, Sophie Bryant, who in 1895 succeeded Miss Buss as Head of the North London Collegiate School, began her study of mathematics at Bedford during the late 1860s.

[7] One of the teachers at the Ladies' College, Lucy Harrison, later set up the Gower Street School. Among its students were Ethel Oliver, daughter of Daniel Oliver, and her friend Edith Chick, the future wife of Arthur Tansley.

[8] www.royalholloway.ac.uk/aboutus/ourhistory.

[9] A graduate of North London Collegiate School, Raisin became a fellow of the Linnean Society in 1906.

with a view to the construction of new buildings, offering much improved facilities, and unfettered growth.

The architect of the new buildings was Basil Champneys, who also happened to have been the architect of Newnham College, Cambridge, opened in 1875. Perhaps not surprisingly, Bedford and Newnham (the latter was originally known as Newnham Hall) displayed a notable similarity from the outside. In 1913, Bedford moved to its new premises for which the Treasurer and now Chairman of Council, Major Leonard Darwin, had been busy raising funds, telling potential supporters that the buildings and their setting would be 'without exaggeration the most charming that could have been selected in London'. The words are taken from a public appeal Darwin was making, through the pages of *The Spectator* magazine (22 October 1910), for £20,000 to match the £20,000 already provided by London County Council. He noted that, 'Great difficulty is now experienced in raising money for University Colleges, and the higher education of women suffers especially from this charge, because no ancient endowments exist for this purpose'. The same sentiment was repeated by many others in the first years of the twentieth century.

University College, London, was in 1826 the first university to be founded in England since the Middle Ages but significantly, unlike Oxford and Cambridge universities, it was open to non-conformists and it was more open-minded with respect to the education of women.[10] In the session of 1861–1862, John Marshall delivered at 3 o'clock on Saturday afternoons a course of 13 lectures on animal physiology to a class composed exclusively of women, 113 of them. The course cost one guinea (£1.05), though there was a 50% reduction for 'ladies involved in Education' (Hale Bellot 1929, 24). Under the auspices of a London LEA, University College provided from 1868 a range of lecture courses for women, these mirroring similar initiatives in Edinburgh and several towns in Lancashire and Yorkshire, as noted previously, where staff of local tertiary colleges (often future universities) provided courses for women under the umbrella of a local LEA. The final step in this particular pathway was taken in 1878 when the University of London formally admitted female undergraduates onto its degree courses, with the first two women to be

[10] The four Scottish universities, St Andrews, Glasgow, Aberdeen and Edinburgh, were open to non-conformists. They did not admit women to degree courses until 1892. In 1836 UCL joined with King's College (founded 1829) to form the new University of London.

awarded a B.Sc. degree graduating in 1881. The Faculty of Medicine was an exception; women were not given equal status until 1917.

A spate of new building was prompted around 1878, not only to accommodate the influx of women but with the aim of transforming the natural sciences into practical subjects, with hands-on teaching in a laboratory (fieldwork was to follow only several years later, see below). However, it was not until 1880, that the University of London introduced a practical component to its examinations in selected subjects. D.H. Scott was made responsible laboratory classes in botany from 1882 to 1885, following a decision that that subject should finally include a practical element (Anon. 1927).

To take a step back, and return to Miss Buss and Miss Beale, it was in 1865 that they joined with Emily Davies, Elizabeth Garret Anderson (née Garrett), Barbara Bodichon (née Leigh Smith) and others in forming the Kensington Society, a small group which met quarterly to discuss what was required, and how it might be achieved, covering issues ranging from women's education to suffrage.[11] The Kensington Group was in practice a meeting of suffragists, though this term was only clearly defined from 1903 when Emmeline Pankhurst formed the break-away Women's Social and Political Union (the suffragettes) which advocated much more militant tactics (Chap. 1).

One of the first tangible outcomes of the Kensington Group's meetings was the founding of Girton College in 1869; originally located in Hitchin, it moved four years later to the village of Girton on the northern outskirts of Cambridge. Its first Principal was Emily Davies, while Barbara Bodichon was a major benefactor (her various benefactions amounted to over £16,000; equivalent to well over a million pounds today). Playing a key role in the design of the gardens during the 1870s was Gertrude Jekyll, the soon-to-be renowned horticulturalist (Chap. 9) and old friend and travelling companion of Bodichon (Walker 2006, 125).

The organisation that was to be Newnham College followed shortly, in 1871, thanks to the inspiration and tireless efforts of the Cambridge philosopher Henry Sidgwick. Among Newnham's most ardent and vocal supporters was the young and newly married Millicent Fawcett (née Garrett), who would later emerge as the leader of the suffragist movement. Her

[11] It was Bodichon's address, 'Reasons for the Enfranchisement of Women', given in Lydia Becker's home city of Manchester in 1866, which inspired Becker's interest in suffrage (Crawford 1999, 42–46).

home in Cambridge was the first meeting place for the founders of Newnham, and a place where supporters of women's education met regularly. In 1876, Sidgwick married one of Newnham's first students, Eleanor Balfour, a sister of one his own students, Arthur J Balfour, the future Prime Minister. Eleanor (*aka* 'Nora') was an exceptionally able mathematician who for a while assisted her brother-in-law, the future Lord Rayleigh (John William Strutt), with experiments to find more accurate ways of measuring electrical resistance (Gould 1997).[12] Married and fully committed to Sidgwick's philosophy, Eleanor later turned her attentions to the college. Although she filled the role of Newnham's mathematics tutor, she was first its vice-principal (1880) and then, after her husband died, its principal from 1892 until 1900, finding endless ways of using her family wealth to support college activities.

In Science: The Laboratory

If women were to make a serious contribution to original research in the natural sciences, which some women agreed with the commonly held male view was a necessary prerequisite to their enjoying full access to learned societies, they would require not only training in the basic skills of the laboratory (maybe field work too) but, subsequently, a place at a laboratory bench and access to costly equipment. Only a few, select, institutions offered laboratory places to women.

The Cavendish laboratory in Cambridge opened in 1874 to teach heat, light, electricity, and magnetism to students of mathematics, but it was not until 1882 that its classes were opened to women. The change was thanks in no small part to the enthusiasm of the Director of the Laboratory, Lord Rayleigh, who believed that women were the equals of men. When Rayleigh was replaced by J.J. Thomson, he was replaced by a man who believed, in direct contrast, that women were not the equals of men, convinced that they could not cope with the advanced classes in physics. Nevertheless, undeterred and guided by Eleanor Sidgwick, Newnham appointed tutors in physics, such as Helen Klaassen, who helped maintain a trickle of female 'graduates' in that subject (Fig. 3.1). A colleague of Klaassen's at Newnham was Ida Freund; it was her job to prepare students of chemistry for Part 1 of the Tripos examinations for, until they had

[12] Strutt married Evelyn Balfour.

Fig. 3.1 Newnham College teaching staff, 1896. Back row: Helen Klaassen (second from left); front row: Ida Freund (third from left); Eleanor Sidgwick (fifth), Margaret Tuke (sixth), and Philippa Fawcett (far right). (Source: Reproduced by courtesy of the Principal and Fellows of Newnham College, Cambridge)

passed that examination, the women were excluded from the University Chemistry Laboratory.

The richly endowed Davy-Faraday laboratory of the Royal Institution, opened in 1896 by Edward, Prince of Wales, was in theory open to researchers of pure and physical chemistry of both sexes. In practice, it offered little to women for, as noted by Kathleen Lonsdale FRS and one of the few women to work in the Laboratory, it was not until 1923, when Sir William Bragg became its Director, that the proportion of women on the staff of the Davy-Faraday approached 20% (Dyhouse 1995, 144).

For aspiring young female botanists of the period, such as Helen Klaassen's sister, Henderina, the situation was somewhat better. The opening of Kew's Jodrell Laboratory in 1876 and the Balfour Laboratory for Women, in Cambridge, in 1881, could hardly have been more timely for the development of their careers. The Jodrell owed its beginning not to government but to a private benefaction, of £1500, from T.J. Phillips Jodrell, a philanthropist and good friend of Sir Joseph Hooker, the Director of Kew. A modest single-storey building was constructed with, most importantly, two well equipped laboratories, one for chemical work, the other for microscopy. Management was invested first in the hands of William Thiselton-Dyer, Hooker's son-in-law and Assistant Director, and then in 1892 it passed to D.H. Scott who was 'Keeper' of the laboratory for the next 14 years (Metcalfe 1942). In the early years, there were no paid staff and visitors had to pay their own expenses. Nevertheless, the Jodrell attracted a stream of the ablest researchers, some already distinguished, like John Lubbock, others who would be so in the future, like Sydney Vines. Many of these younger men had had recent experience of the German laboratories and were inspired by the 'New Botany', with its emphasis on evidence gained from experimentation, which they had witnessed there under the tutelage of men like Julius von Sachs and Anton de Bary.

The Jodrell was open to both sexes and it was here, working under Scott, that Ethel Sargant laid the foundations of her career and for a while collaborated with his wife, Henderina. The first woman to publish a research paper bearing the Jodrell's address—a seminal event—was Sara Agnes Calvert, and her paper appeared in 1887. It described collaborative work with Leonard Alfred Boodle, demonstrator at the Royal College of Science (Calvert and Boodle 1887). Calvert was, however, the sole author of a second paper, published in the same year (Calvert 1887). Both papers were in volume one of an exciting new journal, the *Annals of Botany*, inspired in part by the 'New Botany'. Among the editors of the *Annals* were Scott, Thiselton-Dyer, and Vines (Jackson 2015). Little is known of Calvert: according to Mary Creese she studied at the Royal College of Science 'in the 1880s and 1890s'—where she was probably one of Scott's students—and she obtained a B.Sc. in 1895 (Creese 1998, 60). Thereafter she disappeared from the botanical scene, while Boodle became an FLS in 1888 and, after working in South Africa, was Scott's assistant at the Jodrell from 1904.

Given the difficulties faced by women wishing to study the natural sciences, it was a remarkable event when a biological laboratory specifically for women was opened in Cambridge in 1884 (Fig. 3.2). Three years previously the University had agreed to allow women to sit the Tripos examinations (though not until 1948 could they receive formal degrees)

Fig. 3.2 The Balfour Biological Laboratory for Women, University of Cambridge. The bust of Francis Maitland Balfour overlooks the students' worktables. The laboratory was housed in what had formerly been a Congregational chapel. (Source: Reproduced by courtesy of the Principal and Fellows of Newnham College, Cambridge)

and the purpose of the 'Balfour Biological Laboratory for Women' was to prepare students of the women's colleges, Newnham and Girton, for those examinations.[13]

The Laboratory's origins were pragmatic as well as idealistic. It helped solve the problem of overcrowded mixed-sex practical classes. When Michael Foster had taken the chair of physiology in Cambridge, in 1873, he had enthusiastically begun introducing the practical, laboratory-based teaching which at the Royal College of Science had been a feature of the methods of his erstwhile master, Thomas Henry Huxley. With support from the young embryologist and morphologist, Francis Maitland Balfour

[13] Somerville Hall and Lady Margaret Hall opened in Oxford in 1879 but women could not formally graduate until 1920.

(whose sister, Eleanor, married Henry Sidgwick), Foster's classes grew not just in scope but in size, their popularity attributable not just to Foster's charisma but to an expansion of the medical school, and also the demand from increasing numbers of would-be teachers of zoology in schools. In overcrowded laboratories, it was the women who had found themselves quite literally squeezed to the side, and often ignored. It was a story repeated when Sydney Vines introduced practical classes in Cambridge's Botany School (Richmond 1997, 578). (In later years, after Vines had taken the Sherardian Chair in Oxford, one of his potential students, Ethel Wallace, wrote of her 'dismay and disappointment' when she found that 'old professor Vines was strongly opposed to women students and would tolerate only women who were quite certain of a First Class [degree]' (Griffin 1986, 73). She opted for zoology.)

In the Balfour Laboratory (which was named to commemorate the early death of the brilliant young F.M. Balfour in a climbing accident in 1882, and whose building was half-funded by his brother Arthur J. Balfour), not only could elementary students find the equipment and advice they required, advanced students could carry out their own research projects, and the most able and experienced could find employment as Demonstrators (teaching assistants). The laboratory was housed in an old Congregational Chapel at the heart of the science site in Downing Street and was jointly funded by Newnham and Girton Colleges, which up to that date had been able to offer their students only small, cramped laboratories 'outfitted for work in chemistry' but poorly equipped for biology. The Laboratory soon extended its facilities to include physiological chemistry and physics.

Women who passed through the Balfour Laboratory could regard themselves as members of an exclusive sorority, a feeling reinforced by the Natural Science Clubs which sprang up in each college. These clubs gave the younger, less confident women a chance to air their research projects in an unchallenging environment, and gave the more experienced researchers a chance to promote wider discussions, often in joint-college meetings, as for example when Edith Saunders talked about 'The struggle for existence among plants' (1894), or when Ethel Sargant outlined 'Recent work upon the nucleus in plants' (1898). The clubs served to strengthen the spirit of sisterhood and of belonging.

Saunders was a long-serving key member of staff of the Balfour, while Sargant's protégée, Agnes Robertson (see Chap. 5), was later also a member of staff (demonstrator 1911–1914). DH Scott's sister-in-law, Helen

Klaassen, was a demonstrator (1891–1901), albeit not in botany but in physics, after that subject and chemistry were included in the Balfour's teaching.

Edith Saunders was one of several women who formed part of the informal school of the pioneer Mendelian geneticist, William Bateson. Never given facilities for experimental breeding which he considered satisfactory, Bateson managed to make progress in his research by using the Balfour Laboratory and a string of willing collaborators, including his sister, Anna; his sister-in-law, Florence Durham; Dorothea Marryat (future sister-in-law of Gulielma Lister); and Muriel Wheldale. All were Newnham students, although another of the so-called 'Newnham Mendelians', Nora Darwin (his assistant from 1906 to 1910), had never been a member of that college. Bateson was also able to use part of the grounds of Newnham for his experiments (Richmond 2006). His path was smoothed by an appointment made in 1899 by which Edith Saunders, who had been a demonstrator since 1889, was promoted to be Director of the Laboratory, a post she held until the laboratory closed in 1914 when student numbers crashed following the outbreak of WWI. (More will be learned about Saunders in Chap. 8). Thus the Balfour Laboratory had a limited life, unlike the Jodrell which survives to this day as a world-renowned centre for plant research.

In the Edwardian period, there was a third, older laboratory—more accurately, a set of laboratories—significant in the lives of many female natural scientists, and that was at the British Museum (Natural History) (BMNH).[14] The Keeper of Zoology for much of that period was Albert Gunther, President of the Linnean from 1896 to 1900, while the Keeper of Botany was the equally long-serving William Carruthers, a man whose interest in palaeobotany overlapped with DH Scott's and who was President of the Linnean Society from 1886 to 1890. Among the men who supported the candidature of one or more of the first female fellows of the Linnean, were Vernon Blackman, James Britten, Antony Gepp, E Ray Lankester, George Murray, and Alfred Rendle, all of whom were members of the Museum's staff in 1904–1905, or in the immediately preceding years. From the Herbarium at Kew, women's candidature for fellowship was supported by John Gilbert Baker and, his successor as Head

[14] The BMNH separated from the British Museum in 1881 when Natural History's own, new buildings opened in South Kensington. The move from Bloomsbury was completed in 1883.

of the Herbarium, Otto Stapf. It was to the BMNH that DH Scott regularly took his pupils when he was teaching at the nearby Royal College in the years 1885–1892, pupils who included Annie Lorrain Smith and Scott's future wife, Henderina Klaassen. At the Museum they were introduced not only to current research, but to researchers such as Arthur Lister and his daughter Gulielma, with whom Smith struck up a life-long friendship.

Unfortunately, throughout the Edwardian period and beyond there was continuous tension between the BMNH and Kew as the two establishments vied to take advantage of the great research opportunities which the British Empire was providing in the form of newly discovered plants, and animals. 'The whole energy of Kew, and of the British Museum, of Edinburgh and Glasgow... has been concentrated on the floristic exploitation of the British dependencies', observed FO Bower, sometime worker at Kew's Jodrell Laboratory and later Professor of Botany at the University of Glasgow (Morton and Noble 1983). The crux of the problem was that there were two national herbaria, one at Kew, the other at the BMNH, each wanting the lion's share of the new material and each jealously protected by its staff. Into the arena stepped the government.

While Britain was enjoying the wealth that derived from its dependencies, Germany was building its wealth in another manner, through scientific research and industrial power. Unification of the German States in 1871 helped cement the close collaboration between that country's universities and those manufacturing industries which underpinned its industrial success. Among many steps taken by a concerned British government, aiming to strengthen the nation's competitiveness, was the establishment of a 'Commission of Scientific Instruction and the Advancement of Science' (the Devonshire Commission). Its final report (1884) included the recommendation that 'opportunities for the pursuit of investigations in Physiological Botany should be afforded in the Royal Gardens at Kew' (Metcalfe 1942). It was a recognition that proper exploitation of the botanical riches of the dependencies would require more than the mere description and classification of newly discovered plants.

If that was bad news for the BMNH, worse was to come. Throughout the 1880s and into the 1890s, relations between those two proud institutions festered, with, for example, Kew accusing the Museum of neglecting the long-established practice of sharing duplicate specimens. The matter was not resolved until Her Majesty's Treasury, which had a stake in both institutions, set up a 'Departmental Committee on Botanic Work', which took evidence from all interested parties. In the words of Ray Desmond, who has chronicled the history of Kew, its combative Director, William Thiselton-Dyer, made by far the 'largest and most exhaustive submission' in which he 'extolled Kew as a place of research, dismissing the BMNH as merely a repository'; it was a submission in which Dyer also made a personal attack on James Britten, acting Director of the Botany department at BMNH (Desmond 1995, 286–287). Much to the chagrin of the BMNH and its staff, when the Committee published its report in 1901 it recommended that the two herbaria should be merged—at Kew.

The staff of the BMNH need not have worried. Nothing had happened by 1904–1905, when the Linnean was considering making female fellows, nor need they have been worried in 1906 when, at the BAAS meeting in York, the President of Section K (Botany), Professor Francis Oliver of UCL, used part of his Presidential Address to make once again the argument that the herbaria should be merged—at Kew.[15] Oliver further criticised the sterility of the systematic work of the BMNH, which he believed was divorced from new developments in botany happening elsewhere. He argued that staff of the BMNH should have teaching posts in universities, such as his own department at UCL, which would help them integrate their studies with what was going on not just in the universities but in laboratories throughout the country, such as The Jodrell (Anon. 1907, 733–738). In other words, he implied that they needed to catch up with the world of 1906.

Such rationalisation never happened in the radical way some intended. Compromise prevailed as over the years the BMNH and Kew tacitly agreed to work symbiotically, with each developing its own specialisms. Personal relations between most individuals at the two institutions were good, although the strained professional relations at a senior level must inevitably have impinged on the lives of the more junior staff, and the women working with them.

In Science: Field Work

In mixed laboratory classes, limited space and equipment could often be prioritised for male students. In the field, women were disadvantaged in another way: because of their clothing.

[15] Francis Wall Oliver was the son of Daniel Oliver.

Gwen Raverat (née Darwin) and her cousin Nora Darwin were both born in 1885. As teenagers and young adults they were acutely aware of Edwardian fashions. Gwen remembered that

> *Except for the most small-waisted, naturally dumb-bell-shaped females, the ladies never seemed at ease …their dresses were always made too tight, and the bodices wrinkled laterally from the strain*
>
> *…enormous over-trimmed hats [which] were fixed to the armature of one's puffed-out hair by long and murderous pins.* (Raverat 1954, 222)

As the journalist and social commentator Charlotte Humphry confessed in her book, *Manners for Women*, 'Women dress irrationally… No one knows so well as women how very inconvenient modern dress is' (Humphry 1897, 61).

In passing, it is amusing to note that the subject of women's attire managed to invade even the sober, male-dominated meetings of the British Association for the Advancement of Science (Green 2013, 63–65). It was Charlotte Stopes, mother of Marie, who managed to hijack two of the annual BAAS meetings. While a classics teacher at Cheltenham College (1877–1879), Charlotte had developed an interest in geology (Green 2013, 36). It was geology which had originally attracted her to meetings of the BAAS, but it was not geology *per se* that was on her mind in 1888 when, during the meeting in Bath, she ascended the platform after one session of programmed speeches was finishing (a session which included Lydia Becker arguing for the importance of comfortable bodices) and proceeded to address the audience about the evils of tight-lacing. At the next year's meeting in Newcastle, Charlotte and her supporters went one step further. They organised their own formal session, with an appointed chair, in defiance of the official Programme Committee. Their targets were 'savage attire' and 'the tyranny of fashion' and their particular audience was the press—which they successfully reached.

If the long dresses, heavy coats, boots, and large hats, which were fashionable for everyday life in the late-Victorian/early-Edwardian age, were unsuitable for field work on a fine day, then on a more typical cold, damp, and muddy British day they were both burdensome and imprisoning. When the wind blew, hats were like 'mighty sails'. Although some designers were beginning to produce clothing specifically for increasingly popular outdoor activities, such as boaters (lightweight straw hats) and divided skirts (*culottes* or 'bloomers') for cycling enthusiasts, or more radical innovations for the privileged few who indulged in alpine climbing, most women simply had to put up with the inconvenience of clothes that were completely unsuitable for strenuous field work.

Women embarking on field work suffered a further impediment, especially if they were young and single—as were most students. They could not venture into the countryside on their own; propriety demanded that they should be closely chaperoned, ideally by a mature, married woman. And field work in the presence of young men was often ruled completely out of order. Thus, in the 1870s, when Archibald Geicke, Professor of Geology at Edinburgh University, took his mixed parties of students on field trips, which might last several days, they were accompanied by his wife and another older woman. Geicke pioneered such field trips in the autumn of 1872 and they became a regular feature of his teaching. When in the summer of 1891 Maria Ogilvie Gordon (Chap. 6) was taken on a month-long field trip in the South Tyrol by her professor, Ferdinand von Richthofen, his wife accompanied them as Maria's chaperone. At the all-women Bedford College, Catherine Raisin, head of geology and botany, was reputed to adopt the simple strategy of marching her geology students away whenever men were sighted during field work (Burek and Kolbl-Ebert 2007).

It seems that the geologists were years ahead of the biologists in introducing women to field work. Cambridge's exclusive [Adam] Sedgwick Club, which was founded in 1880 and is the oldest student-run geology society in the world, opened its doors to women in 1896, when it increased the limit on its number of men from ten to twelve and allowed up to six women members. Among its activities, the Club held regular field excursions (Fig. 3.3). When the botanist Arthur Tansley—who would distinguish himself as the man who would shape ecology in Britain—was a student at Cambridge, his first experience of lengthy field excursions was in geology. As a second year student, in 1892, he joined one of the two-week-long courses led by Professor Thomas McKenny Hughes, a man who was known as a strong supporter of equality for women and was someone who fought alongside Geicke and others for women to be admitted to the Geological Society. The courses were mixed, so Tansley could not only learn geology at first hand from a gifted teacher but he could also, in theory, enjoy the company of female students from Newnham and Girton Colleges. In practice, there was not much time for flirtation because the professor worked his students hard and his young wife, Mary, *aka* 'Carrie', 'a wise and excellent chaperone', was a constant presence

Fig. 3.3 Setting off on a geological expedition to the Lake District, July 1890. Thomas McKenny Hughes is seated, front left; his wife, Carrie, stands near right (both wear 'deer-stalker' hats). An anonymous correspondent of *The Queen and Lady's Newspaper* (2nd August 1890), observed that while the young women arrived 'anaemic and nervous', they left 'rosy and vigorous' after twelve days in wind and rain. (Source: Courtesy of The Sedgwick Museum of Earth Sciences, University of Cambridge)

protecting the virtue of the young women in her charge (Godwin 1957). More than that, Mary Caroline Hughes (née Weston) soon earned the respect of those female students in another way, as a thoroughly competent geologist and palaeontologist, publishing papers in her own name (Anon. 1917).

The educational value of such field work was not lost on Tansley. In his first post-graduate job, as assistant to Professor Francis ('Frank') Wall Oliver, at University College, London, Tansley was an enthusiastic supporter when his professor proposed in 1903 to take some botany students to see his research site at Blakeney on the Norfolk coast. The successful excursion, during which students helped Oliver with measurements and undertook simple experiments, was repeated. Soon, more ambitious trips were undertaken, to Bouche d'Erquy on the northern coast of Brittany (Oliver 1906). Parties included both sexes: indeed, in the plans for the 1905 expedition the team from UCL was split into three groups, the one which Tansley led consisted of five women and no men.[16] Intriguingly, the first woman named is a 'Miss Robertson'. No first name is given. However, since Agnes Robertson (of whom there will be more later) was a teaching assistant at UCL at this time, it was in all probability her. Uncertainty remains because the same leaflet states, 'Miss Janet Robertson has kindly consented to make a series of drawings illustrative of the vegetation'. Janet Robertson was a well-known and accomplished botanical artist and the leaflet seems to distinguish two women, both present on the trip. Turning to Oliver's group, it was comprised of four men and only one woman, but she was a certain Miss [Marie] Stopes. Thus, probably present on the 1905 expedition were two women destined to become the most celebrated female botanists of first decades of the twentieth century.

And When the Women 'Graduated'?

The most common occupation of the women who took paid work after 'graduating' from Newnham and similar colleges was school teaching, a career that was clearly encouraged for, in December 1877, Newnham Hall hosted the first-ever national conference of schoolmistresses (Sutherland 2006, 102). Of 371 women who entered Newnham and Girton in the years up to 1880, 230 went into teaching at some level, though very, very few in higher education (Sutherland 2006, 109).

In 1897, a quarter of a century after Newnham was founded, its Principal, Eleanor Sidgwick reflected that among the students of her college, and of Girton, there were women 'who study for study's sake; who desire to devote their time to learning, or science, and hope, however humbly, to help carry knowledge a little further…Yet there are fewer facilities for them in later life; fewer advanced studentships, fellowships and professorships' (Sidgwick 1897). As non-matriculated students, women in Cambridge were barred from university competitions for prizes and fellowships (Richmond 2006, 578).

[16] Anonymous leaflet. *Botanical Expedition to the Bouche d'Erquy. September, 1905.* Tansley Archives, University of Cambridge Library.

Almost identical thoughts were expressed by Ethel Sargant (Chap. 6) in her contribution to the *Jubilee Magazine* of Miss Buss' North London Collegiate School. After bemoaning the lack of professorships for women, she pointed out that the few post-graduate studentships which existed for women, 'rarely last more than two or three years, so that just as the holder of one becomes able to see her way to the heart of her subject she is compelled to stop working at it' (Sargant 1900, 6).

Landing a university position was extremely hard; for example, by 1910, nearly forty years since its opening, there had been only fourteen female researchers in Cambridge's Cavendish Laboratory (Fara 2018, 128). Where teaching was included in a job description there was for female applicants an additional problem if classes included men—a potential lack of respect. As experience taught Elizabeth Morley, the first female professor in an English University college (Reading), the woman needed to be naturally assertive, any timidity on their part giving appointing committees an easy 'get out' as they chose to overlook female applicants (Dyhouse 1995, 139). Non-discrimination was however sometimes practised in the opposite direction, as when, famously, George Barger was appointed in 1913 to the Chair of Chemistry in the all-women Royal Holloway College.[17]

An honourable exception to bias against women was to be found in the Botany Department of University College, London. During F.W. Oliver's tenure of the Quain Professorship (1890–1927), the College's prestigious Quain Studentship was held by eight people, four of whom were women: Edith Chick, who married Arthur Tansley; Agnes Robertson, a future FLS *and* FRS; Sarah Martha Baker, who died young but not before becoming an FLS in 1914; and, in the very last years of Oliver's tenure, Violet Anderson. Perhaps even more importantly, Oliver found staff appointments in those same years for Ethel Thomas, Marie Stopes, and Winifred Smith (all mentioned in the pages which follow) plus a Miss B Russell-Wells and a Miss GL Naylor (Anon. 1927). As will be seen further in Chap. 8, Frank Oliver was, like D.H. Scott, a key figure in the advancement of young women seeking to build a career in botany.

Women's education in general, and in the natural sciences in particular, made huge advances in the second half of the nineteenth century. As the

[17] Queen Victoria opened the buildings of Royal Holloway College, Egham, Surrey, in 1886. In 1900 it became part of the University of London, and in 1985 merged with Bedford College.

twentieth century opened, progression beyond a first degree was still extremely difficult, but not impossible, for ambitious women scientists. Career opportunities were still not keeping pace with the growth in numbers of women educated in the natural sciences.

References

Anonymous. 1907. *Report of the Seventy-Sixth Meeting of the British Association for the Advancement of Science, York, 1906*. London: John Murray.

———. 1917. Obituary. Thomas McKenny Hughes MA. *Geological Magazine* 4: 334–335.

———. 1927. *An Outline of the History of the Botanical Department of University College, London*, 11–12. London: University Press. Issued by the Department on the Occasion of the Centenary of the College.

———. 1969. Schools: Queen's College, Harley Street. In *A History of the County of Middlesex: Volume 1, Physique, Archaeology, Domesday, Ecclesiastical Organization, the Jews, Religious Houses, Education of Working Classes to 1870, Private Education from Sixteenth Century*, ed. J.S. Cockburn, H.P.F. King, and K.G.T. McDonnell. Oxford: University Press for the Institute of Historical Research.

Bailey, Marilyn Ogilvie. 2000. Obligatory Amateurs: Annie Maunder (1868–1947) and British Women Astronomers at the Dawn of Professional Astronomy. *British Journal for the History of Science* 33: 67–84.

Burek, Cynthia V. 2007. The Role of Women in Geological Higher Education. Bedford College, London (Catherine Raisin) and Newnham College, Cambridge, UK. In *The Role of Women in the History of Geology*, ed. Cynthia V. Burek and Betty Higgs, 9–38. The Geological Society of London Special Publication, 281.

Burek, Cynthia V., and M. Kolbl-Ebert. 2007. The Historical Problems of Travel for Women Undertaking Geological Fieldwork. In *The Role of Women in the History of Geology*, ed. Cynthia V. Burek and Betty Higgs, 115–122. The Geological Society of London Special Publication, 281.

Buxton, Meriel. 2008. *The High-Flying Duchess. Mary Du Caurroy Bedford 1865–1937*. Leicester: Woodperry Books.

Calvert, Agnes. 1887. The Lactiferous Tissue in the Stem of *Hevea brasiliensis*. *Annals of Botany* 1: 75–77.

Calvert, Agnes, and L.A. Boodle. 1887. On Lactiferous Tissue in the Pith of *Manihot glaziovii*, and on the Presence of Nuclei in This Tissue. *Annals of Botany* 1: 55–62.

Crawford, Elizabeth. 1999. *The Women's Suffrage Movement: A Reference Guide 1866–1928*. London: University College Press.

Creese, Mary. 1998. *Ladies in the Laboratory? American and British Women in Science, 1800–1900*. London: Scarecrow Press.

Desmond, R. 1995. *Kew: The History of the Royal Botanic Gardens*. London: Harvill Press.

Dyhouse, Carol. 1981. *Girls Growing Up in Victorian and Edwardian England*. London: Routledge and Keegan Paul.

———. 1995. *No Distinction of Sex? Women in British Universities, 1870–1939*. UCL Press.

Fara, Patricia. 2018. *A Lab of One's Own. Science and Suffrage in the First World War*. Oxford: Oxford University Press.

GIanquitto, Tina. 2013. Botanical Smuts and Hermaphrodites: Lydia Becker, Darwin's Botany and Educational Reform. *Isis* 104: 250–277.

Godwin, H. 1957. Arthur George Tansley, 1871–1955. *Biographical Memoirs of Fellows of the Royal Society* 3: 227–246.

Gould, Paula. 1997. Women and the Culture of University Physics in Late Nineteenth Century Cambridge. *British Journal for the History of Science* 30: 127–149.

Green, Stephanie. 2013. *The Public Lives of Charlotte and Marie Stopes*. London: Pickering & Chatto.

Griffin, Penny. 1986. *St Hugh's: One Hundred Years of Women's Education in Oxford*. London: Palgrave Macmillan.

Hale Bellot, H. 1929. *University College, London, 1826–1926*. London: University Press.

Hughes, Mary V. 1978a. *A London Girl of the 1880s*. Oxford: University Press.

———. 1978b. *A London Home in the 1890s*. Oxford: University Press.

Humphry, Charlotte E. 1897. *Manners for Women*. London: Ward, Lock & Co.

Jackson, M.B. 2015. One Hundred and Twenty-Five Years of the *Annals of Botany*. Part I. The First 50 Years (1887–1936). *Annals of Botany* 115: 1–18.

Lockyer, N. 1869. Lectures to Ladies. *Nature* 1: 45–46.

———. 1870. The Scientific Education of Women. *Nature* 2: 117–118.

Metcalfe, C.R. 1942. A Short History of the Jodrell Laboratory. *Chronica Botanica* 7: 174–176.

Morton, A.G., and Mary Noble. 1983. Botany and Mycology. *Proceedings of the Royal Society of Edinburgh, B* 84: 65–83.

Oliver, F.W. 1906. The Bouche d'Erquy in 1906. *New Phytologist* 5: 189–195.

Raikes, Elizabeth. 1908. *Dorothea Beale of Cheltenham*. London: Archibald Constable.

Rappaport, Helen. 2001. *Encyclopaedia of Women Social Reformers*. Frances Mary Buss. Oxford: ABC-Clio.

Raverat, Gwen. 1954. *Period Piece. A Cambridge Childhood*. London: Faber and Faber.

Rayner-Canham, Marelene F., and G.W. Rayner-Canham. 2008. *Chemistry Was Their Life*. In *Pioneering British Women Chemists, 1880–1949*. London: Imperial College Press.

Richmond, Marsha L. 1997. 'A Lab of One's Own'. The Balfour Biological Laboratory for Women at Cambridge University, 1884–1914. *Isis* 88: 422–455.

———. 2006. The 'Domestication' of Heredity: The Familial Organisation of Geneticists at Cambridge University, 1895–1910. *Journal of the History of Biology* 39: 565–605.

Ruskin, J. 1865. *Sesame and Lilies*. London: John Wiley.

Sargant, Ethel. 1900. Women and Original Research. *Frances Mary Buss Schools Jubilee Magazine*, November.

Sidgwick, Eleanor. 1897. *University Education of Women. A Lecture Delivered at University College, Liverpool, May 1896*. 26pp. Cambridge: Macmillan & Bowes.

Sutherland, Gill. 2006. *Faith, Duty and the Power of Mind. The Cloughs and Their Circle, 1820–1960*. Cambridge: University Press.

Walker, Lynne. 2006. Women Patron-Builders in Britain: Identity, Difference and Memory in Spatial and Material Culture. In *Local/Global. Women Artists in the Nineteenth Century*, ed. Deborah Cherry and Janice Helland, 121–136. Ashgate: Colchester.

CHAPTER 4

How Mrs Farquharson Triumphed but Was Excluded from a Glittering Occasion

The key figure in women's gaining admittance to the Linnean Society was Mrs Farquharson; without her nothing might have happened for decades. Her life and struggles with the Linnean illustrate wider contemporary attitudes among both men and women towards women's place not just in the institutions of science but in the wider world. She was fortunate in having family connections to lower ranks of the aristocracy and to 'the establishment', which she used to her advantage, but other women who had similar connections sat back and did nothing. What they lacked was her energy, resolute nature, and willingness to challenge the *status quo*.

It remains a mystery whether it was simply the effrontery of her challenge, or her personality and social attitudes, which upset the members of the Linnean, causing them in 1904 to reject her alone from among the first sixteen female candidates for fellowship. In a photograph taken of her in the same year, she has the appearance of everyone's favourite 'granny', small, warm, and loveable (Fig. 4.1). She was much more than that: she was an enigma.

Descended from a Protestant Martyr

Born on 2 July 1846 at West Meon, Hampshire, Marian Sarah Ridley was the eldest daughter of the Rev. Nicholas James Ridley and was a descendent of the Protestant martyr, Bishop Nicholas Ridley, about whom she

P. Ayres, *Women and the Natural Sciences in Edwardian Britain*, Palgrave Studies in the History of Science and Technology, https://doi.org/10.1007/978-3-030-46600-8_4

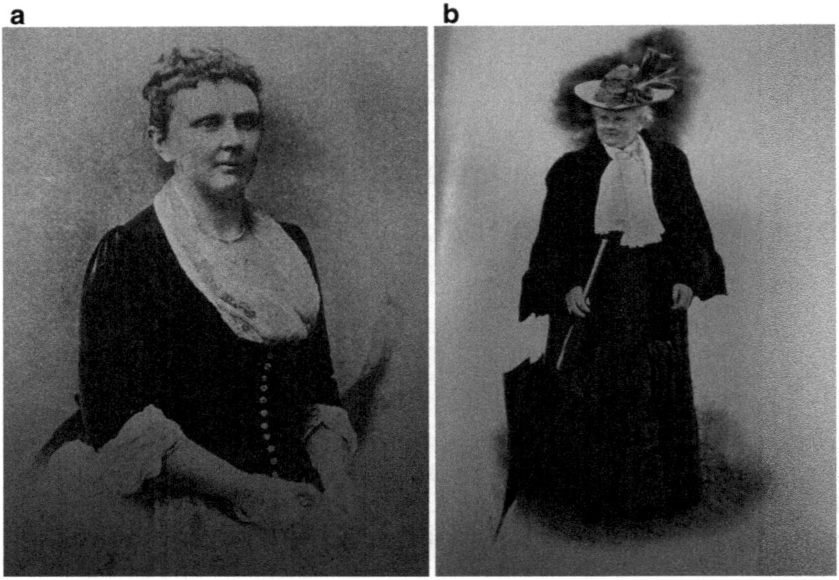

Fig. 4.1 Mrs Robert Farquharson (née Marian Ridley), (**a**) during her marriage and sometime before 1898, (**b**) *ca.* 1903–1904, during widowhood at Tillydrine House. Her attempts to join the Linnean and other societies were made in the years between the photos. (Source: Fig. 4.1a is from Fraser-Mackintosh (1898), Fig. 4.1b is from Royle (1903))

wrote admiringly, he 'preferred to suffer death by burning rather than shirk the courage of his opinions' (Douglas 2010).[1]

Her father was vicar of St Thomas' church in the parish of East Woodhay, near Newbury, Berkshire, but he was no penurious clergyman. His grandfather was the 2nd Baronet Ridley. The Rev. Nicholas James was able to pay for his seven sons to be educated at leading public schools, such as Eton or Winchester. Four went on to Oxford University and two to Cambridge. Nine servants were employed at the large family home, Hollington House, and the family also kept a smart London address, 7 Cambridge Square, Hyde Park. Indeed, Marian may have been living at the latter when she 'attended classes in London' following some earlier

[1] Also: https://doi.org/10.1093/ref:odnb/55777.

private education at home (Douglas 2010). She and her two sisters never experienced the schools or universities enjoyed by their brothers.

In an interview given in 1904 to a journalist, Alice Royle of the *Boudoir Magazine*, Marian admitted that her early thoughts of leaving home to study music, her first love, had been stifled by ill health (Royle 1903). Botany was another interest, and a healthy one, which she could more readily pursue from home, probably encouraged by her father who, in 1870, was listed as a member of the newly formed Newbury and District Field Club (NDFC), though there is no record of him having been among the many active 'parson naturalists' of the age.[2] An invaluable starting point for any botanical interest was certainly provided by a paper in volume one of the *Transactions* (1870–1871) of the NDFC, by Henry Reeks, FLS. It was entitled, 'A list of the flowering plants, ferns, and mosses observed in the parish of East Woodhay'.

Marian was by 1881 able to publish *A Pocket Guide to British Ferns* in which she acknowledged the help of Mr Britten of the British Museum and Dr Baker of Kew, both FLS, and of Dr Murie and Mr West at the Library of the Linnean Society (Ridley 1881). She felt keenly that her work would have benefitted if she had had [direct] 'access to the Herbarium of Linnaeus', adding '...although I was told my election would have been easily carried, it could not be on account of my sex' (Farquharson 1900, 184–185).[3] On October 29th of the same year, however, she was elected a member of the Essex Field Club, in May 1882 presenting to that young club's growing reference library an album of thirty eight herbarium sheets of British ferns (Anon. 1882).

Her life changed fundamentally two years later when she married a Scottish laird, Robert Francis Ogilvie Farquharson (RFOF). She was thirty-seven years old, he was sixty and had two daughters by a previous marriage. His Haughton estate at Alford, 25 miles north-west of Aberdeen, extended to 4500 acres. He was a deputy Lord Lieutenant of Aberdeenshire. Their wedding was at St John's church, Paddington, London, with a reception at 7 Cambridge Square, among the more significant of the fifty guests being Sir Thomas and Lady Farrer, and Sir Arthur and Lady Hobhouse (née Farrer). The connection to the Farrers was through

[2] Founded in 1870, the first President and four of the seven Vice-Presidents were 'Rev.' gentlemen; one of them being Charles Kingsley. Only one woman is listed among the 109 ordinary members.

[3] Marian's paper was read in her absence by Lady Marjorie Gordon, daughter of Lady Aberdeen.

Marian's grandfather, the Rev. Henry Colbourne Ridley, a barrister, who had married Mary Farrer, daughter of James Farrer, a lawyer who was building an extremely successful practice in Lincoln's Inn Fields, London. The fact that both Sir Thomas and Sir Arthur were at the reception strongly suggests that as well as a kinship there was a friendship involving shared interests and ideals. Where Sir Thomas was concerned, there were shared botanical interests. He was a keen and able botanist, his wife was a niece of Charles Darwin, and their home at Dorking, Surrey, was a place where not only Charles and his son Francis visited, but so too did eminent botanists such as Hugo De Vries from Amsterdam (Van der Pas 1970).

Sir Thomas and Sir Arthur, both in their mid-sixties at the time of Marian's wedding, were 'establishment' figures, senior members of an important Liberal network that, among others, included Lord Avebury (formerly Sir John Lubbock), the influential banker, politician, biologist, and both friend and neighbour of Charles Darwin. Lubbock, it will be seen, was a sponsor of Marian's application(s) to join the Linnean Society.

Lord Farrer FLS served as Vice-Chairman of London County Council when Avebury was its Chairman. Letters between the men reveal an intimate friendship. Lubbock worked with Lord Hobhouse, lawyer and judge, on many matters, including the Endowed Schools Act (1871). Hobhouse's interests included reformation of the property rights of married women; in 1870 he published an influential pamphlet, *On the forfeiture of property by married women* (Manchester: Alexander Ireland).[4] He was also the 'Arbitrator' appointed by parliament under the Epping Forest Act of 1878, inevitably becoming acquainted with the officers and activities of the Essex Field Club.[5]

After marriage, Marian moved to her husband's estate, thereafter signing her letters and articles, 'Mrs. Farquharson of Haughton' and adding 'FRMS' (Fellow of the Royal Microscopical Society). The first part, in particular, might have been seen as an irritating affectation and, in widowhood, a title she had no right to use.

[4] Available online. As noted in Chap. 1, it was not until the Married Women's Property Act of 1882 that a woman's wealth, property, and income ceased passing to her husband on marriage.

[5] The Commons Preservation Society (CPS) stopped private landlords enclosing parts of the forest. Epping's future was vested in the Corporation of the City of London. William James Farrer, brother of Sir Thomas (created baronet 1883; 1st Baron Farrer 1893), married the sister of George John Shaw-Lefevre, barrister and founder of the CPS.

Her married years, 1883–1890, were devoted largely to her husband and his estate. Nevertheless, she joined the Alford Field Club and East of Scotland Union of Naturalists' Societies, and it was not long (1885–1886) before she was publishing 'Notes on mosses of the north of Scotland' (Farquharson 1885–1886)—mosses were a new interest—and then a lengthier paper, 'Ferns and mosses of the Alford district' (Farquharson 1889–1890). In September 1885 she presented a paper, 'The identification of British mosses', to the annual meeting of the British Association for the Advancement of Science, held that year in Aberdeen. She shared with her husband an interest in desmids, unicellular green algae found in freshwater, *Docidium farquharsonii* being named in their honour by John Roy after they had collected it from Tent's Moor, St Andrews. In a paper on the desmids of a Hampshire lake, Roy wrote that he was 'greatly indebted' to Mr Farquharson and 'his accomplished wife…for collections of Desmid material' (Roy 1890). He explained that 'For the last few years they had to pass the winter in the South of England', a further indicator of the ill health which was increasingly to mar Marian's life.

After seven years of happy marriage, her husband died following a bout of influenza.[6] As well as losing her husband, she lost her home, Haughton House, and all its happy associations, for ownership passed to her step-daughter, Miss Maria Ogilvie Farquharson (Fraser-Mackintosh 1898, 166–167). Well provided for, and in love with Scotland, Marian moved to Meigle, in Perthshire, and later to Tillydrine House, a large Italianate house overlooking the River Dee and Grampian mountains, at Kincardine O'Neil and only sixteen miles from Alford (Royle 1903).

Possibly as a distraction from the grief and loneliness of widowhood, she revived her interest in algae. The archives of the Linnean Society include letters written over the next three years to Edward Morrell Holmes FLS, an acknowledged expert on that group—and also on mosses.

Neither socially nor scientifically were there grounds for Marian's ultimate rejection by the Linnean. She was related to the baronetcy, and she had married a major landowner, so socially she was at least the equal of the men deciding upon her candidacy. Scientifically, she was better qualified to be elected a fellow of the Linnean than some women who were successful and whose main qualification was that each was married to an Officer of the Society.

[6] *Aberdeen Journal*, 7 May 1890.

Fighting on Behalf of Women in Science

Marian was a prodigious writer of letters to newspapers, a habit whose importance increased for her as her health and fitness to travel deteriorated. In more than 100 letters to local Aberdeen newspapers, she made frequent reference to papers on the subject of women's work in science which she had presented at meetings organised by various women's groups. There is nowhere, however, any suggestion that she was involved in a broader struggle to improve the lot of lower-class working women. Indeed, by 1909 she had become horrified by the militant activities of Mrs Pankhurst's suffragettes which she felt were harming the wider cause of women's equality (Pedersen 2004, 35). Although she claimed as a 'friend' Dr Millicent Fawcett [sic], President of the peaceable National Union of Women's Suffrage Societies (the 'suffragists'), the only evidence of any contact with Fawcett was on the subject of nursing, rather than suffrage, and then it was via the press.[7]

Marian's earliest paper on the position of women in science was prepared for the International Women's Congress of 1890, held in Paris.[8] However, her obituary in the *Aberdeen Daily Journal*[9] reveals that the Paris paper, like others prepared later for the Women's International Congress in London, and for the Glasgow Exhibition, had to be read for her because she was not strong enough to deliver it in person.

Marian's first attempt to join a leading scientific society was when she applied to join the Royal Microscopical Society (RMS). Presumably she felt her interest in desmids made her well qualified to become an Ordinary Fellow. It is not known which of the required three members supported

[7] 'Letters to the Editor', *Aberdeen Free Press*, 2 June, 1904. In a letter printed in the *Times*, 20 July 1908, Marian's name appears next to Fawcett's in a lengthy list. The subject is the registration, or equal professional status, of female nurses.

[8] Cited by Mary R S Creese, 1998, in *Ladies in the Laboratory? American and British Women in Science, 1800–1900*. London: Scarecrow Press, p. 46, also by GL Herries Davies, 2007, in *Whatever in Under the Earth. The Geological Society of London 1807–2007*. The Geological Society of London, but not found by this author. Another paper, 'Women's Suffrage: Should it be Made a Test Question?' is wrongly attributed to Marian in the *Biographical Dictionary of Women in Science*, 2000, edited by Marian Ogilvie Bailey and Joy Dorothy Harvey, London: Routledge, p. 436. The published introduction to that paper, delivered to a meeting held in Dundee in 1899, states that it was given in person by a Mrs Farquharson who was the daughter of the ex-Provost of Dundee, Alexander Mathewson.

[9] Death of Mrs Farquharson of Haughton. A notable career. *Aberdeen Daily Journal*, 13 May 1912.

her application but page 568 of the *Proceedings* of the RMS record that at a meeting of 8 April 1885 she was admitted. She was not, as has sometimes been stated, the Society's first female fellow for the *Report to Council, 1884*, noted that four ladies had already been made fellows, among them Catherine Crisp, wife of Frank Crisp, Treasurer of the Linnean. And membership was only a partial success for, until 1909, women were not allowed to attend meetings of the Society or to take part in its discussions. Thus, for a quarter of a century, they did not enjoy *full* membership, an important distinction.

It is less clear why Marian's interests should have led her towards the Anthropological Institute (AI), nevertheless the records of the AI for 1900 show that she was among its fellows. Perhaps it was simply the act of joining as many societies as possible that was important to her for she allowed this particular fellowship to lapse after only two years. The AI's Chair of Council was the same Lord Avebury who was to sponsor her application to join the Linnean Society. Marian also challenged the Royal Geographical Society, something which, as will shortly be seen, met with the severe disapproval of one eminent female geographer. Again, Avebury was involved and, in this case, so too was Lord Aberdeen (another distant relation by marriage).

In a *Life of John Lubbock*, written in 1914, only a year after his death, his biographer says, 'Sir John never sympathised with the exclusion of women from scientific societies, and spoke in favour of their admission to the Geographical Society'. An excerpt from a letter written to Avebury by Sir M.E. Grant Duff, past-President of the Society, and dated 17 April 1893, is then quoted (Hutchinson 1914, 17):

> *One of those idiotic squabbles which now and then disturb learned Societies has broken out in the Geographical. The subject...is the admission of women to be Fellows....*

Anticipating a future meeting, Grant Duff adds: 'Aberdeen will come and talk sense. I hope you will come and help him. With the two all will go well'. However, things did not 'go well' and it was not until 1913 that women were admitted to the Geographical Society.

Emboldened by her association with such an active ally as Lady Warwick (Chap. 1), Marian embarked in 1900 on her battle with the Linnean and Royal Societies, petitioning them both for the admission of suitably qualified women as full fellows. Each letter was accompanied by a cutting from the *Women's Agricultural Times* of November 1899 which reported the resolution of Lady Warwick's Association (Mason 1995).

The response of both societies was negative. As already seen, the Royal Society replied that the eligibility of women, 'must depend on the interpretation to be placed upon the Royal Charters under which the Society has been governed for more than three hundred years'. When three years later the Royal treated Pierre and Marie Curie so unequally, Marian expressed her outrage to the *Aberdeen Free Press*:

> *As your readers are aware, Professor Curie, of Paris, was in regard to radium recently the joint-discoverer with Madam Curie, and had the honour of receiving from the Royal Society on the 30th of November, at the hands of its president, Sir William Huggins, the Society's Davy medal. Sir William gracefully alluding to the name of Madam Curie, who could not be present on account of her sex.*[10]

In passing, it is worth examining Marian's outrage and her interpretation of events in more detail, since they show how her passionate feelings may occasionally have led her to misrepresent the facts. Whereas both the Curies had visited London in June 1903, when Pierre gave an evening lecture at the Royal Institution (a visit during which Marie met Hertha Ayrton for the first time), in November Pierre visited London alone for the simple reason that Marie was unwell, suffering a long-lasting cough which her doctors feared might signal the onset of tuberculosis (Quinn 1996, 185). Pierre stayed with Sir William and Lady Margaret Huggins, a couple who, like the Curies, spent their lives working together although, at the time, Lady Margaret did not receive the credit she deserved for their innovations in spectroscopy which helped lay the foundations of modern astrophysics.

At a later point in the letter above, Marian expressed her hope that the Royal's discrimination against women would

> *...soon be ended by the courtesy and gallantry of the Royal Society following the example of the Linnean Society, which is effecting an alteration in its charter in order to render equal opportunity of Fellowship to both sexes.*

[10] Farquharson, Marian S. Letter to the *Aberdeen Free Press*, 15 December 1903.

Council of the Linnean had placed a series of obstacles in Marian's way, all of which she eventually overcame. In May 1901 she submitted a lengthy list of supporters, whose diverse backgrounds ranged from medicine to the army and included two leading tropical botanists (her cousin, Henry Nicholas Ridley, Director of the Singapore Botanic Garden, and Sir George King, Superintendent of the Royal Botanic Garden, Calcutta). To each rebuff she responded, and each time a new objection was raised (Gage and Stearn 1988, 89–91). She changed her tactics, requesting that she and two friends, Grace Coleridge Frankland and Clara H. Whitmore,[11] be allowed to attend the General Meeting scheduled for November. Permission was not given, in spite of the fact that as recently as June 6th Ethel Barton and Annie Lorrain Smith had been allowed to attend a General Meeting (Gage and Stearn 1988, 90).

With the support of her friends she persisted and won. The proposal for a new Charter was approved by fifty-four votes 'for' to seventeen 'against' on 15 January 1903. Revised bye-laws were approved on 3 November 1904, and on 16 November 1904 the names of sixteen women were presented for election (Marsden 2003).

Those on Her Side

Marian lacked neither direct help nor the support of like-minded associates, whether they were from the natural sciences or from the wider world.

Male Fellows of the Linnean

Each of the sixteen women had to be nominated by a number of fellows, sometimes as few as three, sometimes as many as seven. Marian's 'Certificate' [of Recommendation] was signed by Lords Avebury and Ripon (M.E. Grant Duff), Henry John Elwes, Michael Foster, Joseph Reynolds Green, and William Carmichael McIntosh, each irrefutably a distinguished member of the Society. These six men supported the woman as well as the cause, but why did they approve of Marian when so many others did not?

[11] Whitmore was an American writer who published in 1910 one of the earliest books of feminist criticism, *Women's Works in English Fiction from the Restoration to the mid-Victorian Period* (London: GP Putnam).

It was through Lord Avebury that Marian had submitted her original petition to the Linnean in 1900. Lords Avebury and Ripon were Liberal politicians of influence and both were longstanding members of the Linnean, the former having served as President from 1881–1886. Avebury had reached his seventieth birthday in 1904 and the political and scientific connections which he had built through his long life were virtually endless. Among his friends was the aforementioned Sir M.E. Grant Duff (Lord Ripon), and also Sir Dudley Coutts Marjoribanks (1st Baron Tweedmouth), father of Ishbel, Lady Aberdeen. Whether Marian ever met Avebury is not known, though the Essex Field Club would have provided a natural place for their paths to cross for Avebury was among its founders in 1880 (Thompson 1930). Also, they could have met when they both presented papers in the Biology Section of the BAAS meeting of 1885 in Aberdeen.

After her initial petition to the Linnean, Marian often submitted her requests through Joseph Reynolds Green, who from 1887 was Professor of Botany at the Royal Pharmaceutical Society. Entering Cambridge as a mature student, Green had studied botany under Sydney Vines (President of the Linnean Society in the critical years 1900–1904) and animal morphology under Michael Foster, Professor of Physiology (O'Connor 1991, 29). While Green's botany had recommended him to the Pharmaceutical Society, it was of a 'pure' nature rather than 'applied', so Edward Morrell Holmes was appointed 'Lecturer in *Materia Medica*' to teach students about the medicinal properties of plants, which Green could not (Hudson 2013, 53). There is no evidence that Green knew Marian in person, rather than through the business of Council, but it would have been easy for him to learn about her from his colleague, Holmes. Whatever and however Green knew about Marian, it was enough for him to sign her Certificate.

Another signatory was Foster and again there is a connection to Sydney Vines. Ever since his arrival in Cambridge in 1870, Foster had built both practically and intellectually what became known as the Cambridge School of Physiology (Geison 1978). Foster helped many younger staff, newly appointed zoologists and botanists, to introduce practical teaching into their courses. One of those botanists was Vines. Foster was not only a father-figure for Green and Vines, he was also close to Avebury, succeeding the latter as Liberal MP for the University of London.

William Carmichael McIntosh, zoologist and Professor of Natural History at St Andrews University, was the fifth of Marian's backers. The two were on cordial personal terms. On 1st August 1902, after lunch at

Marian's home, McIntosh had addressed a meeting of the Scottish Association for the Promotion of Women's Public Work chaired by Marian's neighbour and local MP, Robert Farquharson (a potentially confusing name), clearly a movement with which McIntosh had sympathy.[12] Indeed, McIntosh also signed the Certificates, supporting election to the Linnean, of two other women with Scottish connections, Maria Ogilvie-Gordon and Grace Frankland who, from among the sixteen women proposed, were among those most concerned with women's rights (Chaps. 6 and 8, respectively).

Marian's last backer was Henry John Elwes, a botanist, entomologist, and horticulturist. Elwes was a great traveller, often with his friend and brother-in-law Frederick DuCane Godman. He had proven his sympathy with women's involvement in the Linnean's activities when in December 1898 he had introduced two ladies to one of the Society's meetings at which they could hear him 'discourse on the flora and fauna of the Altai mountains', which he had recently visited. At that time, the presence of ladies was not unprecedented but was nevertheless remarkable (Gage and Stearn 1988, 89).

In April 1901, Marian chose to direct her requests to Council via Godman and the Linnean's Zoological Secretary, Thomas GB Howes—presumably chosen because they would be likely to be sympathetic (Gage and Stearn 1988, 90). The first signs that opposition was crumbling followed the meeting of 19 December 1901 when Reynolds Green, in re-submitting Marian's memorial, stated that 'a considerable number of Fellows favoured it' (Gage and Stearn 1988, 90).

Women and the Wider Scientific Community

In November 1900, Marcus Hartog FLS, Professor of Natural History at University College, Cork, had written to the Council of the Linnean urging it to revise its Charter to enable the admission of women. It seems there may have been a link between Hartog and Marian through Hartog's cousin, Hertha Ayrton (née Marks). From the age of nine, when her father died, Hertha had been brought up by her maternal aunt, Mrs Marion Hartog, who ran a small school in London. Here she received a broad, liberal education, alongside her cousins, including Marcus. Her

[12] Archives of the University of St Andrews. ms.37098/26. Letter from Mrs Farquharson to Professor McIntosh, 23 July 1902.

subsequent studies at Girton College, Cambridge, were supported by, among others, Barbara Bodichon and Mary Anne Evans (*aka* George Eliot, the novelist).

A thorn in the side of the older scientific 'establishment', Hertha was, like Marian, outspoken. Unlike Marian, Hertha was a militant, high-profile member of the WSPU working alongside Emmeline Pankhurst and Elizabeth Garrett-Anderson (Mason 1991; Jones 2009). The paths of Marian and Hertha had crossed in 1899 when they were both involved in The International Congress of Women, held in London from 26 June to 7 July. While Ayrton presided over the Physical Science section, delivering a paper, 'The suitability of women for work in the electrical industry', Marian contributed to the biological sciences section. Oddly, although her paper, 'Work for Women in the Biological Sciences', was intended to encourage young women to become scientists, there were some passages which might have had the opposite effect, as when she described her frustrations with the Linnean Society when preparing her book on British ferns. Other female fellows who contributed to the biological sciences section of the Congress were Ethel Sargant and Grace Frankland.

The 1899 Congress was organised under the auspices of the International Council of Women. President of that Council since 1893, the same year her husband was appointed Governor-General of Canada, was Ishbel Hamilton-Gordon, the thirty-six year old wife of the 7th Earl of Aberdeen (Glick 1995, 161). Frustrated by a father's refusal to allow his daughter to receive a university education, Ishbel was a tireless worker, not just an aristocratic figurehead, for a number of women's organisations. Thus, after returning from Canada in 1898 she took on the huge task of editing the seven volumes of proceedings of the International Congress. Marian was distantly related to Lady Aberdeen since that lady's sister, Mary Georgina, was married to Matthew White Ridley, the 1st Viscount (5th Baronet) Ridley.[13] There were, thus, very good reasons why Marian's obituary should include Lady Aberdeen in her list of influential friends and supporters.[14]

Another named 'friend', active in the International Council, and its Corresponding Secretary from 1906—working with the President but relieving her of much routine work—was Dr Maria Matilda Ogilvie

[13] One of Marian's great, great grandfathers was the 2nd Baronet Ridley.

[14] Death of Mrs Farquharson of Haughton. A notable career. *Aberdeen Daily Journal*, 13 May 1912.

Gordon. There will be much more in Chap. 6 about this distinguished and much travelled geologist but what is relevant here is the period in her life when, after marrying John Gordon, an Aberdeen physician, she returned in 1895 to live in Aberdeen, until his death in 1919. She found an outlet for her energies in working for the legal rights of working women, and for improvements in the health of poor mothers and their children. It was probably at about this time that she made the acquaintance of Marian Farquharson for, when a group of women which had coalesced around Marian, and which called itself the 'Women's International Progressive Union', was formalised in 1902 and renamed 'The Scottish Association for the Promotion of Women's Public Work', Marian was its President and bore all its expenses for the first year, while Maria was its Secretary.[15] The Association never flourished and seems to have disappeared with Marian's death, but Maria continued her public works, alongside her geology, in 1919 being among the first cohort of women to be made fellows of the Geological Society.

Maria Ogilvie Gordon, like Grace Frankland, and Ethel Sargant, was included in the first tranche of women to be made fellows of the Linnean Society. They had not been excluded in spite of working for social justice, and in spite of the associations they had with the wider struggle for equal rights for women. Which raises once again the question, if they were admitted, why not Marian?

Those Against Her

Strong-minded and self-confident, Marian laid herself open to accusations of haughtiness, and not knowing 'her place'. Her letters to the Aberdeen newspapers often strayed into what were widely regarded as men's territory. When the re-introduction of tariffs on goods imported into Britain was a political issue, she wrote in defence of free trade. And, backed by 'my ancestor Bishop Ridley', she took on a group of Presbyterian and Church of Scotland ministers on subjects such as baptismal regeneration and the historical Christ (Pedersen 2004, 35).

In botanical matters she could be equally forthright. The abstract of her address to the 1885 meeting of the BAAS finds her criticising Dr Robert

[15] Lindy Moore (pers.comm.), with thanks to [info@women'shistoryscotland.org].

Braithwaite's *British Moss-Flora* which, to modern eyes, is a work of scholarship judged against which her own field guide is lightweight.[16]

> *…no one who commenced the study of this order of Cryptogamia can have failed to experience difficulty in the earlier stage of his [Braithwaite's] work. Without wishing to depreciate the several valuable works on this subject, I have noticed the absence of any work which deals with the distinctive characters of moss, apart from those of a general nature. I feel sure the want is much felt by young students.* (Farquharson 1886, 1063)

Might she then have been seen by her contemporaries as too proud (remember her style of self-address, p. 62), too self-opinionated, or just 'difficult'? Not according to her obituarist who explained her long list of 'friends' as

> *…being due to the ladylike and temperate language of her addresses, as well as to the charm of her personality—the very antithesis of the accepted idea of a "blue stocking" or "new woman"—loveable, kind-hearted, unselfish, tolerant, broad-minded, generous, and sincere….*[17]

Her step-daughter, who inherited the Haughton estate when Marian's husband died, did not entirely agree. She asked readers of the *Aberdeen Free Press* of 11 December 1903 not to hold her responsible for letters signed 'Mrs Farquharson of Haughton':

> *I have to submit to a good deal of inconvenience and annoyance through the sayings and doings of a lady who, in letters to the press, …subscribes herself 'Marian S. Farquharson of Haughton' without having the smallest right to the designation. As my name so closely resembles Mrs Farquharson's (not of Haughton), I find myself sometimes credited with peculiar views about things…which I in no way share.* (Maria O. Farquharson of Haughton)

Also not seemingly charmed by Marian were two significant botanical figures, Dukinfield Henry Scott and Ethel Sargant. It was the job of Scott, Botanical Secretary of the Linnean and a sociable man who naturally enjoyed female company, to receive the paperwork nominating a potential

[16] The first volume of Robert Braithwaite's *Flora* was published in 1887, although some sections were already completed by 1880 and could be purchased separately.

[17] Death of Mrs Farquharson of Haughton. A notable career. *Aberdeen Daily Journal*, 13 May 1912.

fellow. His diary entry for Thursday 15th December 1904 records coldly, '15 ladies elected. Mrs Farquharson rejected, 31 for 19 against'.[18] What is remarkable is that Scott was a man who did not normally bear grudges; he was often the peace-maker in disputes, respected by both sides in an argument (see Chap. 8). Similarly, Ethel Sargant was normally kind-hearted and well-liked; she shared many of Marian's feelings about women's advancement (Chap. 6), and was to be an immediate beneficiary of her efforts, yet she admitted to her young protégée, Agnes Robertson, negative feelings about Marian: 'I couldn't help rejoicing in Mrs F's non-election. It would be disastrous if they felt bound to elect every woman put up'.[19] Ethel could expect a sympathetic hearing from Agnes who only twelve months earlier had written:

> *The Universities have opened their doors to us…The only way in which we can justify the concessions which have been made to us is by doing our share of original work.*[20]

Without referring to Mrs Farquharson as an amateur, the inference is that Sargant thought the latter's contribution to biology was lightweight; though this begs the question of what she thought of the election of the Duchess of Bedford, and the wives of officers. It is remarkable that both Scott and Sargant had no sympathy with Marian's plight.

In making her argument for women fellows, Marian may have made the mistake of aligning herself too much with the great and the good and not enough with rank-and-file fellows. Thus, in the list of twenty-three supporters which she had submitted to Council in May 1901, there were three lords, five knights, eight professors, a reverend, a doctor, and just five with the plain title, 'Mr'.

What sort of men might have been the natural opponents of Marian? We will never know who voted against Marian but the Groves brothers might fit the profile. The sons of a railway clerk, their own careers were modest: Henry as a middle-rank civil servant, James in the Army and Navy

[18] Archives of the Royal Botanic Gardens, Kew.

[19] Archives of Girton College, Cambridge. Letter of 20 December 1904 from Ethel Sargant to Agnes Robertson (who later married Edward Arber).

[20] Arber, A. 1903. The work of women in botanical research. *Biennial Leaflet*. London: The Women's Agricultural and Horticultural International Union. Cited by Schmid, Rudolf. 2001. Agnes Arber, née Robertson (1879–1960): fragments of her life, including her place in biology and in women's studies. *Annals of Botany*, 88, p. 1121.

Stores. Away from their jobs, and always working together—they were popularly known as 'Messrs. Groves'—they developed an unmatched expertise in the Characeae (green algae). According to James, his brother Henry was a Conservative in later years, who 'had little sympathy with the ideals of modern democracy' (Groves 1913). James, a Conservative like his brother, was definitely present (possibly Henry too) at the Special General Meeting held to discuss changing the Charter; James' presence is recorded because he unsuccessfully tried to stop discussion of a second amendment to the Constitution, such that the size of the Council should be enlarged (Pugsley 1933; Marsden 2003).

By attempting to join several societies Marian may in the eyes of some have displayed unwelcome political, rather than intellectual, motivation. And not all women agreed with Marian that their sex should enjoy full fellowship in scientific and learned societies. One such was Mary Kingsley (1862–1900), traveller and explorer of West Africa. When in 1899 a number of women, seemingly led by Marian, made their bid to join the Royal Geographical Society, Mary described the applicants as 'shrieking females and androgynes', refusing to sign Marian's petition (Frank 2005, 256). Using more temperate language, on 26 November 1899, she replied to Marian's request:

> *I feel I cannot add my name to your influential list. I have for many years heard this question about admitting ladies to learned societies discussed and my personal feeling is that I would not ask any Society to admit me. ... If we women distinguish ourselves in Science in sufficiently large numbers at a sufficiently high level, the great scientific societies will admit us...*or... *we will form our own of equal eminence. The great thing for us in this generation to do is to show a good output in high class original work.* (Blunt 1994, 148)

Next day (27 November 1899), she copied to John Scott Keltie, Secretary of the Royal Geographical, her letter to Marian, adding, 'I do not wish to alarm you but I feel it is my duty as a friend to warn you that there is a dangerous female after you, I enclose details. I'm terrified of her' (Blunt 1994, 157–158). Four days later (1 December 1899), she wrote again to Keltie. Citing the Anthropological Institute as an example, Kingsley told him that she found the presence of ladies was 'hindersome to the gentlemen', inhibiting scientific discussion both because of the need for propriety and because of the interests of the ladies (Blunt 1994, 149).

Kingsley was closely associated with Albert Gunther FLS for he had helped to arrange finance for her expeditions. Gunther may have been another example of a fellow who although not against the admission of women in principle—he was a supporter of both the Duchess of Bedford and Catherine Crisp—learned to dislike Marian Farquharson. Gunther was the brother-in-law of William Carmichael McIntosh, whose accounts of Marian's way of life in faraway Aberdeenshire might, inadvertently, have coloured Gunther's opinion of her.

It is most likely that the combination of Marian's character and the effrontery of her challenge was sufficient to prevent enough male fellows from supporting her candidacy.

A Glittering Occasion: Dinner at Princes' Restaurant

> *Today I have the honour of admitting Her Grace the Duchess of Bedford, who was in the first list of Lady-Fellows elected on December 15th. Such additions can be nothing but a strength and an honour to our Society, and the Treasurer has signalised the historic occasion by a dinner to the new Fellows and commissioning a picture of the scene at our meeting on January 19th.*
>
> William Abbott Herdman, President of the Linnean Society, 24 May 1905. (Cited by Savage 1936–1937, 185)

The path to sexual equality in the Linnean had been long and difficult. In his Presidential Address of 1903, Sydney Vines (no great admirer female students: Chap. 3) articulated the anxiety which ran throughout Council when he admitted he was apprehensive about the future, describing the Society as passing through a critical period (Gage and Stearn 1988, 91). Through meeting after meeting in 1903–1904 those in favour of a motion that '…it would increase the usefulness of the Society if women could be elected Fellows…' gradually gained the ascendency and by the autumn of 1904 a revision of the Charter and Bye-laws was finally complete: the new Charter contained the critical words, '…to elect such Persons without distinction of sex to be Fellows…' (Marsden 2003). The way was clear: on 17th November 1904 the names of sixteen women were proposed for fellowships, a ballot was held and at the Society's meeting of 15th December it was announced that fifteen of them had been elected, Mrs Marian Farquharson having been excluded. Eleven of those women were formally admitted, signing the Book of Admission and Obligation, on 19 January

1905, others signing at later dates. To celebrate the admission of the women there was held on 18 May 1905 a splendid dinner at one of London's smartest addresses, Princes' restaurant in Piccadilly, only metres away from the Society's home at Burlington House.[21]

Among marble statues and fashionable potted palms, under golden chandeliers, the glittering occasion was capped when, after toasts to the King, and to the Queen (an Honorary member of the Linnean Society), a toast was offered to the 'Lady-Fellows of the Linnean Society'. A bejewelled Mary Russell, Duchess of Bedford, newly made FLS, gave thanks on behalf of the ladies (Blackwell 1936–1937).

The Duchess was said to have been an ornithologist but on close inspection her interest in birds was probably in its infancy in 1904–1905. She did have though an older interest in biology which would have recommended her to the Linnean. That interest was in the rare deer that roamed the grounds of Woburn Abbey. Mary had been a Fellow of the Zoological Society since 1892 and had worked with the Duke, who was the President of the Zoological Society, to introduce many new species to the park. From 1893 to 1914 she kept a register of the forty-two species in the park, meticulously recording purchases, births, and deaths. The most celebrated species was the rare Père David Deer (*aka* Water Deer, or Milu) which the Duke had rescued from China as they were about to become extinct. The entire world's stock is today derived from the eighteen animals that were conserved at Woburn and it includes descendants of those which the Bedford family was able to re-introduce to China in 1985 (Buxton 2008, 140).

If the Linnean had planned to have among its first 'Lady-Fellows' a woman of the highest rank, it could not have chosen better. This Duchess was more than a figurehead, she was a down-to-earth woman, knowledgeable in two areas of biology—or three if medicine is included (Chap. 9).

Blemished Celebrations

Treasurers of august Societies, such as the Linnean, are expected to be men of the highest probity, combining business acumen and financial expertise with a concern for natural history. Frank Crisp fitted that description perfectly. A wealthy London solicitor, who counted several foreign

[21] The Linnean first occupied part of Burlington House in 1854. In 1873 it moved to rooms newly built onto the original house. It still occupies those rooms today.

railway companies and the Imperial Japanese Navy among his clients, he was generous to the Linnean in many ways (and similarly to the Royal Microscopical Society, of which he was also the Treasurer) (Gage and Stearn 1988, 77–78). Apart from paying for electric light to be brought to the Linnean's rooms at Burlington House, Crisp payed for both the Supplementary Charter of 1903 (Marsden 2003), which eased the way for women to be made fellows, and for a large painting depicting the event on 19th January 1905 at which the first ladies signed the register formally admitting them to the Society. The picture was painted by a distinguished Royal Academician, James Sant. It cost Crisp £300 (roughly equivalent to £30,000 today).

In Sant's picture are seen Miss E.L. Turner, Miss Annie Lorrain Smith, B.D. Jackson, the General Secretary, Miss S.M. Silver, Mrs L.J. Veley, who is signing the book, Mrs Constance Sladen, D.H. Scott, the Botanical Secretary, Mrs Crisp, who is receiving the hand of fellowship from the President, W.A. Herdman, and Frank Crisp, the Treasurer. However, in an older, original picture, which was exhibited at the Royal Academy's Exhibition of 1906, there were two other figures, the Rev. T.R.R. Stebbing, the Zoological Secretary, and his 'somewhat corpulent' wife, who was standing in the right foreground, between Herdman and Crisp. Rev. Stebbing was at the extreme right hand end of the table.

Frank Crisp was far from pleased with the original. Soon after the celebratory dinner, he wrote to Jackson, 'Naturally if I pay £300 for a picture, I should prefer that another Fellow's wife should not be the selected figure!' And on 22nd May he followed his letter with a telegram to Jackson, 'we must surely have at the table a lady fellow who has done something[,] not one without a record[.] glad to assist artistic licence but that would be going too far' (Sheffield 2006). In defence of Mary Anne Stebbing, she was an accomplished water-colourist who specialised in painting British plants (see Chap. 9). After the Academy's Exhibition, the angry Crisp withheld the picture and it was not until shortly after his death in 1919 that Lady Crisp (her husband was knighted in 1907) released the picture to the Society. By that point, both of the Stebbings had been overpainted, Mary Anne having been replaced by an empty chair.

This was not the first occasion on which a Stebbing had been perceived to mar the Society's celebrations. Following the ladies' signing of the Book of Admission on 19th January 1905, the evening's programme allowed for a short, fifteen-minute talk by the Rev. Stebbing on the Crustaceae, which, it was planned, would be followed by a lighthearted address by Augustine Henry FLS on 'Botanical Collecting'. A plant-hunter with a wealth of anecdotes from his travels in China and Indo-China, Augustine Henry (AH) was not only well-known for his sympathy with female emancipation, and a close friend of Henry Elwes FLS, who had been one of the women's main supporters, but AH was an eloquent speaker with an Irishman's fabled gift for words (Pim 1966, 134). As he himself declared, 'It was thought I would be a proper person to enliven the proceedings...' (Pim 1966, 160). In the event Stebbing rambled on for 50 minutes, presenting his audience with preserved crustacean after preserved crustacean, and causing AH to curtail his talk. The mood of the special evening fell decidedly flat.

There was one further blemish on the celebrations of 1905, probably its greatest. That was the absence of Mrs Marian Farquharson, for it was she who had led the struggle for women to be admitted to the Linnean Society. Her treatment was, in the words of the Society's official biographers, Gage and Stearn, 'disgraceful', for she was the 'women's champion' (Gage and Stearn 1988, 91).

An Offer Too Late

Marian's case was finally reviewed in 1908 and she was offered a fellowship. Her Certificate was again signed by Lord Avebury and Joseph Reynolds Green. Edward Morrell Holmes supported her as, fittingly, did three of the women who had been the beneficiaries of her efforts—Catherine Crisp, Grace Frankland, and Ellen Willmott. But Marian's health had deteriorated further. Twice she had to ask Council to postpone the date when she might attend a meeting at which she could sign the Book of Admission that would formalise her fellowship. She visited Nice, on the Mediterranean coast, in the hope that its gentler climate would restore her health. Sadly it did not. A fellowship of the Linnean Society eluded her to the very end for the women's champion died there on 20 April 1912. She was never able to append FLS to her name.

References

Anonymous. 1882. *Essex Field Club Journal* 2: 62.

Arber, Agnes. 1903. The Work of Women in Botanical Research. *Biennial Leaflet.* London: The Women's Agricultural and Horticultural International Union. Cited by Schmid, Rudolf. 2001. Agnes Arber, née Robertson (1879–1960):

Fragments of Her Life, Including Her Place in Biology and in Women's Studies. *Annals of Botany* 88: 1105–1128.

Blackwell, Elsie M. 1936–1937. Obituary. Margaret Jane Benson. *Proceedings of the Linnean Society* 149: 186–189.

Blunt, Alison. 1994. *Travel, Gender and Imperialism. Mary Kingsley and West Africa*. London: The Guilford Press.

Buxton, Meriel. 2008. *The High-Flying Duchess. Mary Du Caurroy Bedford 1865–1937*. Leicester: Woodperry Books.

Douglas, Gina. 2010. Farquharson, Marian Sarah (1846–1912). *Oxford Dictionary of National Biography*. Oxford: University Press.

Farquharson, Marian S. 1885–1886. *Scottish Naturalist* 8: 381.

Farquharson, Mrs., FRMS. 1886. On the Identification of the British Mosses by Their Distinctive Characters. *British Association for the Advancement of Science: Held at Aberdeen in September 1885*. London: John Murray.

Farquharson, Marian S. 1889–1890. *Scottish Naturalist* 10: 193–198.

———. 1900. Women in Professions, 184–185 in the Professional Section of the *International Congress of Women, London, July 1899*, ed. Lady Ishbel Aberdeen. London: T Fisher Unwin.

Frank, Katherine. 2005. *A Voyager Out: The Life of Mary Kingsley*. New York: Tauris Parke Paperbacks.

Fraser-Mackintosh, C. 1898. *An Account of the Confederation of Clan Chattan; Its Kith and Kin*. Glasgow: John Mackay.

Gage, A.T., and W.T. Stearn. 1988. *A Bicentenary History of the Linnean Society of London*. London: Academic Press.

Geison, G.L. 1978. *Michael Foster and the Cambridge School of Physiology*. Princeton, NJ: University Press.

Glick, Daphne. 1995. *The National Council of Women of Great Britain: The First One Hundred Years 1895–1995*. London: National Council of Women of Great Britain.

Groves, J. 1913. Henry Groves (1855–1912). *Journal of Botany* 51: 73–79.

Hudson, Briony. 2013. *The School of Pharmacy, University of London: Medicines, Science, and Society, 1842–1912*. London: Academic Press/Elsevier.

Hutchinson, Horace G. 1914. *Life of Sir John Lubbock. Lord Avebury*. Vol. ii. London: Macmillan.

Jones, Claire G. 2009. Bodies of Controversy? Women and the Royal Society. In *Femininity, Mathematics and Science, 1880–1914*, ed. Claire G. Jones, 175–204. Berlin: Springer.

Marsden, J. 2003. Charter and Bye-Laws. *The Linnean* 19: 13–15.

Mason, Joan. 1991. Hertha Ayrton (1854–1923) and the Admission of Women to the Royal Society of London. *Notes and Records of the Royal Society of London* 45: 201–220.

———. 1995. The Women Fellows' Jubilee. *Notes and Records of the Royal Society of London* 49: 126.
O'Connor, W.J. 1991. *British Physiologists, 1885–1914. A Biographical Dictionary.* Manchester: University Press.
Pedersen, Sarah. 2004. *Within Their Sphere? Correspondence to the Aberdeen Daily Newspapers, 1900–1918.* Ph.D. thesis. The Robert Gordon University, Aberdeen.
Pim, Sheila. 1966. *The Wood and the Trees. A Biography of Augustine Henry.* London: Macdonald.
Pugsley, H.W. 1933. Obituary. James Groves (1858–1933). *Journal of Botany* 71: 136–139.
Quinn, Susan. 1996. *Marie Curie. A Life.* London: Mandarin.
Ridley, Marian S. 1881. *A Pocket Guide to British Ferns.* London: David Bogue.
Roy, J. 1890. Freshwater Algae of Enbridge Lake and Vicinity, Hampshire. *Journal of Botany* 28: 334–338.
Royle, Alice C. 1903. Mrs Farquharson of Haughton FRMS. Botanist and Writer. *Boudoir* 2: 811–816.
Savage, S. 1936–1937. Obituaries. Mary du Caurroy Russell, Her Grace the Duchess of Bedford. *Proceedings of the The Linnean Society* 149: 185.
Sheffield, Suzanne le-May. 2006. Gendered Collaborations: Marrying Art and Science. In *Figuring It Out: Science, Gender and Visual Culture*, ed. Ann Shteir and Bernard Lightman, 240–266. London: Dartmouth College Press.
Thompson, P. 1930. A Short History of the Essex Field Club. *Essex Field Cub Special Memoirs* 7: 2.
Van der Pas, P.W. 1970. The Correspondence of Hugo de Vries and Charles Darwin. *Janus* 57: 173–213.

CHAPTER 5

Miss Sargant and a Botanical Web

In the months between the meeting of the Linnean Society's Council on 3 November 1904 which agreed to alter its bye-laws, thus allowing the Society to offer fellowships to women, and the grand dinner held on 18 May 1905, both to celebrate that decision and to welcome the first women, Council approved for fellowship the names of twenty five women (Appendix). Not all of them had yet signed the Book of Admission and Obligation, whereby the last stage of the formal admission process would have been completed.

There have in the past been attempts to characterise the first women who joined, or attempted to join, scientific societies at around this time. A pattern was recognised among women applying to join the Chemical Society; most of them moved backwards and forwards between a small number of institutions, meeting each other in the process. It was concluded that through such a network the nineteen petitioners of October 1904 became acquainted with each other and able to agree on a common goal. Thirteen of the women overlapped, in twos, or sometimes threes, in the University of London (William Ramsay's group at UCL being one notable mixing pot), while at one time or another there were seven at Newnham College, Cambridge, and four at the University of Bristol—some women being involved in more than one overlap. Prior to university, four of the women had at some time attended King Edward VI High School for Girls in Birmingham, while two had attended Cheltenham Ladies College (Rayner-Canham and Rayner-Canham 2003). Beyond this

P. Ayres, *Women and the Natural Sciences in Edwardian Britain*,
Palgrave Studies in the History of Science and Technology,
https://doi.org/10.1007/978-3-030-46600-8_5

there has been little characterization of the individual chemists, excepting their leaders, such as Ida Smedley.

A more extensive search for common factors among the first twenty-one women who were eventually allowed to join the Geological Society has been made by Cynthia Burek, her examination covering factors such as their educational achievements, place of residence at the time of their election, their age, and their marital status.

> *…no common thread* [was found] *between them, other than their love of geology. It might be assumed that to join a professional organisation like the Geological Society an academic degree was necessary but that does not seem to be the case. Proximity to London and the meetings might have seemed a logical assumption but that does not seem to be the case either with several Fellows being located many miles from the capital e.g. Newcastle upon Tyne. Maria Ogilvie Gordon while listing Aberdeen as her domicile actually had a flat in London.*

Burek (2009), continues:

> *There seems to have been no limit either to the marital status…[and] age was not a barrier with those joining ranging from 27 years (Helen Marguerite [Muir] Wood) to Catherine Raisin at 64 years.*

Turning to the female fellows of the Linnean Society, the greater sample size (25), and, in most cases, the well documented details of their lives, facilitate a comparable search for common factors among these women who sought fellowship with other natural scientists. The Linnean fellows differed from the chemists and geologists in that they were drawn from not just one science but from several, botany, zoology, geology, microbiology, and genetics. Accepting that the analysis may be confounded by those fellows who were not recognisably 'scientists'—who did not publish research papers or books which expanded knowledge of their subject—Chap. 10 will reveal that several characteristics can indeed be recognised among the group. Before that point, however, a detailed exploration must be made of their family background, education, and interests (including those away from the natural sciences). In short, we need to get to know them.

In this chapter, 'getting to know them' begins with five women, all botanists—botany being by far the most widely taught science in girls' schools and colleges—and four of them graduates. All were connected in some way with the first subject, Ethel Sargant; though, as will be seen in later chapters, they were not alone in this.

Ethel Sargant, the Senior Woman

Ethel was one of that small number of girls who had the good fortune to study at Miss Buss' North London Collegiate School. And she was blessed in another respect—with financial freedom. The daughter of a wealthy barrister, she was free to lead the leisured life of a gentlewoman, never having to seek paid employment. She had enough money to build a laboratory where she could indulge her passion, the study of plants, if and when the fancy took her. She was in this sense old fashioned, a throwback to an age when science was commonly conducted by gentlemen in country houses. Ethel was however no dilettante.

Learning About Plants

It is hardly surprising, given her family background, that in her youth she seems to have had a relaxed attitude towards learning; four years after entering Girton College in 1881 (Fig. 5.1), she left with only a third class 'degree' in the final Tripos examinations.[1] She did, however, put her time at Girton to good use in other ways. For example, she stage-managed a production of the *Electra* of Sophocles (her sister, Mary, being recruited to paint the scenery), and she made a host of friends, such as the influential Helena Swanwick who had a small part in the play (Swanwick 1935, 119). These women were both emotionally and practically helpful throughout her life—in which she remained single—just as she was to them.

Ethel may have been slow 'to find herself'—'...home life and some more or less desultory botany occupied [her] for a number of years after she went down from Girton'—but in 1892, for unexplained reasons, her life changed (Arber 1927). She spent most of that year studying in the Jodrell Laboratory, Kew, under the guidance of DH Scott. As she entered her thirties, Ethel Sargant had at last found herself. Her first publication emerged from this period, a joint paper with Scott in the *Annals of Botany*

[1] In 1881, Graces were passed by the University whereby women could officially sit the Tripos examinations for the first time: previously this could only be done by arrangement. However, only 'titular degrees' were awarded by Cambridge until 1947.

Fig. 5.1 First year students at Girton College, Cambridge, 1881. Ethel Sargant is on the left end of the middle row. (Source: Permission of the Mistress and Fellows of Girton College, Cambridge)

(1893), 'On the pitchers of *Dischidia rafflesiana*' (the pitchers are hollow leaves harbouring ants in a mutualistically beneficial relationship).

Her experiences at Kew inspired Ethel to build herself a small laboratory and library, familiarly called 'Jodrell Junior', in the grounds of her mother's house, 'Quarry Hill', at Reigate, Surrey, and it was there that her best work was done. She was, where botanical research was concerned, in control of her own destiny. She had the luxury of being able to take things slowly, letting her ideas mature. It was an approach to planning research which she strongly recommended to others.

The beauty of having a laboratory 'at home' meant that Ethel could manage the increasing demands of her ageing mother and of her dear sister, Maudie, 'whose mind had never fully developed, and who had to be cared for like a young child' (Arber 1927). The needs of these two always had a higher priority in Ethel's life than did botany, a situation with which her two other sisters and four brothers were happy to go along. Frustratingly, family duties never allowed Ethel as much time as she needed and wanted to devote to her personal research, yet she kept her investigations moving steadily forward, not least because she regularly employed at least one assistant.

The purpose of the laboratory in the Reigate garden was a mystery to the neighbours, and also to the local excise man. He paid frequent and surprise visits in the belief—strengthened by his observation of apparatus for distilling water—that the laboratory's true purpose was the production of illicit alcohol. In 1909, Ethel left Junior Jodrell behind, moving her household to Calverley Park, in Tunbridge Wells.[2] Only three years later, in 1912, she moved on when, following the deaths of both Maudie and their mother in close succession, Ethel chose her last home. It was at the Old Rectory in Girton village, and there she established another laboratory though her most productive years were over. She was in familiar territory, full of happy memories. Her old college made her a fellow in 1913 (Robinson and Perry 2008).

Ethel's studies centred on the plant cell and it was only later in life, when her eyesight deteriorated, that she turned to the study of tissues rather than individual cells (Arber 1919). Between 1895 and 1900 she published six cytological papers dealing with the nuclei of the embryo sac, and their fertilisation by male nuclei, in *Lilium martagon* (Turk's Cap lily). Incidentally, she had great difficulty in persuading her mother's gardener to grow the lilies in large blocks, 'like vegetables', rather than in small, visually pleasing clumps (Arber 1927). It was during this period that she and a fellow botanist, Margaret Benson, undertook a tour of the continent, finding 'Botany is certainly more thought of in Germany than here, and a Professor in his own institute is a very great man indeed'.[3] Although Ethel made much progress in revealing the processes of oogenesis and spermatogenesis, to her chagrin she missed something important, the 'triple fusion' of the second male (pollen) nucleus with the two polar nuclei in the egg sac. She was scooped by a Russian and a French researcher; when their results became known in Britain in 1899, she looked back at her discarded microscope slides and was able to confirm their observations, extending their conclusions to 'her' species, *L. martagon*. She had been, in her own words, 'An Ass' (Arber 1927).

[2] Letter to Agnes Arber (née Robertson), 3 October 1909, Archives of Girton College, Cambridge.

[3] Letter from Ethel Sargant to Agnes Arber, 28 Nov 1897. Archives of Girton College, Cambridge.

While at the Jodrell, she had first made friends with DH Scott's wife, Rina. Together they investigated the development of seedlings of wild *Arum* (a tuberous perennial, often, wrongly, called a lily). An anecdote about their collaboration exemplifies Sargant's sense of humour, which all her friends recognised but which is hidden in every photograph and drawing of her.

> *Mrs Scott and I actually read our* Arum *paper at the Botanical Club on Tuesday, and it went off extremely well.*

It went so well that

> *…an animated discussion arose which ended in the joint authors contradicting each other vigorously! We had omitted to compare notes on the subject of root hairs, and Mrs Scott thought them present on the roots of* Arum*—I that they were absent. I believe she was right.*[4]

The women remained lifelong friends. The *Arum* work paved the way for the second phase of Ethel Sargant's research career, which she called the years of 'anatomical morphology'. In this phase she studied the transition zone between root and shoot, where the anatomy of a plant, particularly its vascular system, changes fundamentally. Whereas most studies to date had dealt with the transition zone in dicotyledonous species, Ethel concentrated on monocotyledons, where the transition zone is much smaller and more difficult to resolve.[5] These studies were facilitated by her exceptional ability to cut thin ultra-thin sections (slices) of plant material: her protégée, Agnes Arber (née Robertson), believed that Ethel had learned the basic technique from Cambridge's zoologists, thus making her the first botanist to use the new Cambridge Rocking Microtome, which automatically cut an ultrathin ribbon of sections of biological material embedded in wax. The sections could then be stained to differentiate the various tissues.

There were two main outcomes from this period. Arising directly from her collaboration with Rina, Ethel spent many years unravelling the mechanisms by which the contractile roots of monocotyledonous species help

[4] Letter Ethel Sargant to Agnes Robertson, 16 June 1898. Archives of Girton College, Cambridge.

[5] Seeds of flowering plants produce either one or two seedling leaves (cotyledons). Grasses, lilies, and orchids are examples of the first type, roses, daisies, and beans of the second.

pull the seedlings down into the soil. In 1911 she stayed with the Thomas Hanburys at La Mortola, their famous gardens on the Italian Riviera, where she was able to make comparative studies (mostly unpublished) of their extensive collection of monocots, particularly from the Iridaceae (Arber 1919). On a different subject, and working with her assistant Ethel Thomas—a woman who was to build a successful career as a botanical researcher for herself—Sargant had by 1902 concluded that the single cotyledon of monocots was homologous with *both* cotyledons of the dicots; it was simply in a state of fusion. As she refined her theory, over the next decade and more, she concluded that monocots were derived from dicots, an idea that was soon being widely discussed and tested by the botanical community, though not always accepted.

Ethel's botanical interests were not confined to laboratory studies of the anatomy of living or fossil plants. She was an active member of the Holmesdale Natural History Club (Chap. 3) and was from 1898 to 1905 its Treasurer. One of Ethel's many interests was orchids. The records of the Essex Field Club show that in 1903 she wrote from the Holmesdale to the Essex's members asking for information about orchids in their region, but appealing to them not to dig up clumps of these relatively rare plants (Anon. 1903).

Family, Suffrage, and Politics

Throughout her life Ethel was torn between her love of plants and her love for her family, and it was family responsibilities that usually dominated. She felt a responsibility for her sister Maudie which only intensified as their ailing mother grew older. Ethel was quite prepared to sacrifice her own welfare for Maudie's benefit, as exemplified by two letters to Agnes Robertson written in the winter of 1900:

> February 13th
>
> *Maudie is no better: this snow has brought on her rheumatism again and all my time is taken up in nursing her. Unfortunately her maid, though most willing and patient, is no nurse.*
>
> March 7th
>
> *Maudie is getting better: we have got a proper sick nurse for her who is a great success. My mind is relieved, and besides I have some time to myself.*[6]

[6] Letters Ethel Sargant to Agnes Robertson. Archives of Girton College, Cambridge.

Ethel's shouldering of domestic responsibilities left her siblings free pursue their own lives. And what distinguished people they were! One brother, Charles, became a high court judge, another, Walter, 'the acclaimed headmaster' of Oakham School. Her brother Edmund was essentially an educational reformer, at one time teaching at Toynbee Hall, though he never settled for long in one job and spent a long time in various capacities in South Africa. There was a further brother, Francis, who was a sculptor. There was also a sister, Alice, who published a small book, *The Crystal Ball. A Child's Book of Fairy Ballads* (1894), illustrated by her sister, Mary. Alice's death in 1909 was a significant event for not only did it persuade Ethel to close Junior Jodrell, sell the house in Reigate and move to Tunbridge Wells, but a bequest made by Alice to Bedford College, her *alma mater*, formed part of a set of unhappy events described in Chap. 8. Excepting Maudie, probably the most significant sibling in Ethel's life was her older sister, Mary, who was both an artist and a suffragist.

After studying in Paris and then at the Slade School of Fine Art in London, Mary had married an American musician, Henry Smythe Florence, in 1888. Tragically, Henry drowned just three years later. By 1905 Mary was listed as a subscriber to the Central Society for Women's Suffrage, and it is known that next year, on 11 December, she was present at a banquet at London's Savoy Hotel organised by Millicent Fawcett, President of the NUWSS, to celebrate the release from prison of members of the rival organisation, the WSPU. Mary became a subscriber to both the NUWSS and the WSPU (Crawford 1999, 223). She became a leading light of the Tax Resistance League—which advocated no taxation without representation—not only employing her artistic talents to design their badge and several banners but also withholding her taxes in 1912 and 1914, leading to her goods being distrained and sold.[7] As a dedicated pacifist, as well as a feminist, Mary was a stalwart of the Cambridge' Heretics' Society and, with the group's founder, Charles Kay Ogden, was co-author of the book *Militarism versus Feminism*: London, Allen & Unwin, 1915 (Forrester and Cameron 2017, 109).

Ethel's own enthusiasm for the cause of women's suffrage probably owed much to her older sister's example, although she may also have been influenced by friends she made while at Girton; friends such as Helena Swanwick (née Sickert), a member of the NUWSS from 1906 and editor

[7] https://womanandhersphere.com/2013/06/.

of its journal, *The Common Cause*, from 1909 to 1912.[8] Whatever the origins of Ethel's enthusiasm, she could find encouragement locally because the Tunbridge Wells branch of the NUWSS was one of the society's most active (Crawford 1999, 690). In 1910, thanks to financial support from the ever-generous Ethel, her local branch was able to rent a shop in Tunbridge Wells where meetings and fund-raising 'teas' could be held, and from which literature could be sent out (Fig. 5.2). The shop's window was filled with posters, including some created by her sister Mary. For the election of the same year, Ethel hired a motor car which, decked in Liberal party colours, was used by party officials to tour the area and mobilise Liberal voters—something which, her mother felt, was 'doing Ethel all the good in the world and throwing her with young and interesting women'.[9] Such activity proved an antidote to the sadness following the recent death of her 'darling Maudie'.

'It is polling day here', Ethel wrote to Agnes Robertson in December 1910, '…our Suffrage Society is very busy. Am going to sit at the receipt of customers in the shop this afternoon. All this last paragraph is in a whisper, you need not hear it unless you choose'.[10] Ethel's caution was because she knew Agnes did not support the suffrage movement, in spite of her own sister, Margaret, being a leading activist.

Sensitive to their disagreement over suffrage, Ethel had explained her own position to Agnes at some length in a letter of 11 September 1908:

> *I was very struck by a remark made by a man* [who] … *struck me as a Conservative…He said that the question of the future…was the family and that it would be disastrous if women had no direct share in settling it.*
>
> *I think the whole position of women to be in a state of unstable equilibrium. If it does not advance it must go back.*
>
> *I should not hesitate to give it* [my casting vote] *to the enfranchisement of women on the same terms (property qualification) as those on which men now hold it. For adult suffrage I do not think the times are ripe.*

[8] Helena was the sister of the artist Walter Sickert, who had studied briefly at The Slade (1881) and thereafter in Paris before settling in the artists' community in Chelsea (see Chap. 8).

[9] 13 March 1911, a letter from Ethel's mother to Mary: cited in Crawford (1999, 690). Polling in General Elections took place on different days in different parts of Britain and Ireland.

[10] Letter to Agnes Arber (née Robertson), 7 December 1910. Archives of Girton College, Cambridge.

Fig. 5.2 The Suffrage Shop in Tunbridge Wells, Kent, a centre for fund raising, *ca.* 1910. (Source: https://www.flickr.com/photos/lselibrary/40080806642/in/photolist-244NBqo)

What Ethel Sargant wanted in politics, as in science, was the equal treatment of men and women. She was not campaigning for full adult suffrage. Although her involvement with suffrage came relatively late in her life, she had long before set out her attitude to the position of women in science in articles written for the *Girton Review* (1901) and for the *Jubilee Magazine* of North London Collegiate School (1900).

> *The truth is we do not realise the experimental phase is past. Women have shown that they can stand University tests and endure the same mental discipline as men.*
>
> *Our graduates [from Women's colleges] can point with pride to their achievements in the Finals.* (Sargant 1901)

She felt that in the natural sciences women were disadvantaged in several ways. Not principally because they may be 'less given to independent action than men, [but] chiefly I believe because their early training has made them more conscientious and less enterprising'. In addition, unless they enjoyed independent means—and here she must have recognised her own good fortune—there were few scholarships, and even fewer professorships, open to women who needed to support themselves financially. Foreshadowing her feelings about suffrage, Ethel felt that women's position in science was also in a state of unstable equilibrium. The time was ripe for them to assume a greater role. She felt keenly the exclusion of women from learned societies.

> *Custom and some degree of prejudice close the doors of most of the learned societies against women. In that respect we are little better off than our predecessors of fifty years ago. But when we meet on neutral ground I have always found the attitude of scientific men generous to women workers. We cannot but regret that we are so often excluded from the stimulus of comradeship.* (Sargant 1900, 8)

The same words could have come from Marian Farquharson's pen.

Things would have to change if the advances of the last fifty years were not to be wasted. Now was the time for women to prove their worth by doing 'original research' which would contribute significantly to the corpus of scientific knowledge.

Protégées

Ethel did everything she could to promote the careers of young women in botany. For example, at the Junior Jodrell she employed Ethel Thomas as her assistant from 1897 to 1901 and, when Ethel Thomas left, her place was taken by Agnes Robertson for the year (1902–1903) between the completion of her studies at Newnham College and the commencement of her postgraduate studies at UCL. This was far from the first time Agnes had been at Reigate. She had visited briefly during her final year at school (1897), and while an undergraduate at UCL (1897–1899) she had spent time there in at least one summer vacation, learning techniques for microscopy.

The exceptionally close relationship between Sargant and Robertson, 15 years her junior, is evident in the many letters exchanged between the two women and now held in the archives of Girton College. At an early stage in the friendship, Sargant visited Robertson's parents, with whom a warm relationship developed. When the younger woman was preparing to apply for admission to study natural sciences in Cambridge, but undecided about the relative virtues of the two women's colleges, Sargant was uniquely qualified to give advice. She wrote:

All my prejudices are in favour of Girton as a college so this advice goes against the grain. Science however is abominably neglected at Girton, they have very few science students and the laboratory is a howling wilderness. At Newnham there are numbers of science students, and all the best choose to go there…The laboratory is beautifully fitted up…[11]

Among the Newnham staff upon whom Sargant showered praise were Miss Klaassen (Rina Scott's older sister), for physics, and Miss [Becky] Saunders, for botany. Miss Robertson followed the advice proffered to her: she chose Newnham.

The relationship between the two women was always conducted in a most circumspect manner; it was not until June 1907 that 'Miss Sargant' proposed they should thereafter address each other by their first names. As Robertson embarked on her own career of research and publication, she sent drafts of her first papers to her older friend for comments and suggestions.

Ethel Sargant was made a member of the Council of the Linnean Society in May 1906, the first woman to be so honoured. Although she valued highly her fellowship, and membership of Council, she viewed proceedings however with her usual mix of irreverence and humour.

I attended my first Council meeting last Thursday and as you foretold the atmosphere was pure 18th century, that is the moral or emotional atmosphere. The physical air can hardly date from earlier than 1870 or so, and intellectually we were quite late in the XIXth century.

Continuing:

…immediately after the minutes had been read I was startled by Dr Scott performing a lightening act of disappearance—more like Harlequin at the Pantomime than a steady going scientific gent. The next business however explained his action: it was estimates [of printing costs] *for his forthcoming paper and the Referee's report on it.*[12]

When later that year Robertson announced her engagement to the palaeobotanist Edward Arber, Sargant proposed Robertson for a 'Life Fellowship

[11] Letter from Ethel Sargant to Agnes Robertson, 13 March 1898. Archives of Girton College, Cambridge.

[12] Letter from Ethel Sargant to Mrs. Agnes Arber (née Robertson), 10 June 1906. Archives of Girton College, Cambridge.

of the Linnean Society', in spite of both women's evident reservations. It was to be a wedding present; Ethel Thomas was offered a similar fellowship, but as what Sargant jokingly called an 'unwedding present'. In January 1909, seven months before her marriage, Agnes Robertson was admitted as a Fellow of the Linnean.

Edward Arber died after the couple had been married for only nine years, leaving Agnes alone to bring up their daughter Muriel—a future geologist of distinction—and also to build her own career in botany, which she did so successfully that in 1947 she was made a Fellow of the Royal Society, the first female botanist ever to receive that distinction. Agnes' career bore all the hallmarks of her mentoring by Ethel Sargant. As Ethel had taught her, Agnes believed that researchers should not involve themselves in teaching, for it only dulled their originality. After the Balfour Laboratory closed, Agnes set up her own laboratory, just as Ethel had done, and there in seclusion, in Cambridge, she relentlessly pursued her own interests, keeping herself at arm's length from the university and its botanists (Schmid 2001). One obituarist credited Agnes Arber with being both a 'botanist and philosopher', for her writings were integrative, drawing on a wide range of sources from Wolfgang Goethe (who wrote widely about the ancestral forms of plants) to Spinoza, and often including a sprinkling of metaphysics (Stearn 1960).

Ethel Sargant was treated by the female botanists of her generation as the 'senior woman', even if she was not the oldest of them. They held her in the highest esteem, looking to her for guidance and leadership. She was often their point of contact with the leading male botanists, for those men too held her in the greatest respect.

Margaret Jane Benson, Professor, Royal Holloway College

We had Dr Benson—Dr Bunny as she was commonly called. Well, she was just a dear. For all her learning there was a most loveable naiveté about her. She was a staunch teetotaller, and nobody had the heart to tell her why the Sunday evening trifle tasted so particularly nice. (Miss W. E. Delp, Head of Department of German, Royal Holloway College, 1908–1944[13])

[13] W.E Delp. 'Royal Holloway College, 1908–1914'. Archives of Royal Holloway College, University of London, RHC/131/6. Published online to celebrate International Women's Day 2017. The College had been opened by Queen Victoria in 1886.

When Margaret Benson was in 1912 made a professor, almost twenty years after joining the staff of Royal Holloway, a women's college of the University of London, she was the first female Professor of Botany in the country, and was one of only a handful of women who at the time held a chair in a British university. As Miss Delp suggests, she seems to have personified the popular image of the brilliant but unworldly academic (Fig. 5.3).

Margaret was introduced to botany by her father who, when he was not working as an architect and civil engineer, was able to indulge his interest in field botany, returning from every expedition into the countryside with a full sketchbook. Her love of drawing may have been inherited from her father or, alternatively, from her mother, Edmunda (née Bourne), a talented artist who at the age of seventeen had exhibited a flower painting at the Royal Academy.

Fig. 5.3 Margaret Benson, Head of the Botany Department at Royal Holloway College, Egham, Surrey. (Source: Supplied from the archives (PP26/10/7) of Royal Holloway College, University of London)

Margaret was the sixth child in a family of nine. All that is known about her early education is that for a period she was taught at home by her elder sister, Henrietta, who had once attended Queen's College.[14] Her later education was a patchwork of institutions and subjects. At the age of nineteen (1878–1879), she spent a short time studying classics at Norwich House, Cambridge, which was associated with Newnham College (Blackwell 1936–1937). The Register of Students for Bedford College records that she attended classes in Mathematics, Botany, English Language, German, Latin, History, Drawing and Music from Michaelmas 1879 to Michaelmas 1880. Her next step was to take a post as an assistant mistress at Exeter High School. Whether it was always her intention to study botany is not known but, by saving money from her teaching salary, she was able by 1887 to register at University College, London (UCL) to study science. In 1891 she was awarded a B.Sc. degree.

An outstanding student, she won a Marion Kennedy research studentship to stay on at UCL studying the embryology of the Amentiferae (an artificial group of catkin-bearing, wind-pollinated trees) under the supervision of Professor F.W. Oliver. At that point in his career, Oliver's interest was turning to fossil plants, an interest encouraged by his friendship with the palaeobotanist D.H. Scott. Their interest rubbed off on Margaret for it was fossil plants that were to fill her future life in research, first during a brief spell at Newnham (1892–1893), and then, beginning the same year, 1893, at Royal Holloway College after her appointment as the Head of its new Department of Botany. In 1894 she was awarded a D.Sc. by the University of London.

One of Margaret's first projects at Royal Holloway was to create a 'botany garden' in the College's extensive grounds at Egham, Surrey. Ever since the sixteenth century, when such gardens provided the *materia medica* for teaching student doctors, gardens formed a vital part of any course of botanical study—they were often the only practical element. Margaret clearly wanted more than a teaching aid for her aim was to assemble there a collection of hardy perennials for research purposes. She was quite prepared to do much of the spadework herself! Hoping to amass more than seventy examples, and on the advice of Professor Harry Marshall Ward (Cambridge), she asked William Thiselton Dyer, Director of the Royal Botanic Gardens, Kew, for help. Whether the misogynistic Dyer instructed

[14]The schoolbook of 'Maggie J Benson' (aged 12), with its charming drawings of plants and animals, is among the archives of RHC.

his staff to send specimens to her is not recorded. However, the man who was both Dyer's predecessor at Kew and his father-in-law, Joseph Hooker, certainly did donate plants to Margaret's garden. Five years after arriving at the College she was able to report that her garden had been enlarged to such an extent that Evelyn Welsford, a gardener trained at the Duchess of Warwick's Studley Horticultural College, had been employed to oversee it (Blackwell 1936–1937). Her perseverance had paid off; perseverance which, she had defined in an essay written at age twelve, was '…the prime quality of life. Men fail much oftener from the want of patient perseverance than from the want of talent' (Blackwell 1936–1937).

Just as her D.Sc. work had focussed on the reproductive structures of extant flowering plants so, when she turned to palaeobotany, she again focussed on reproductive structures—in this case of the pteridosperms (seed ferns) and lycophytes (club mosses and other spore bearing vascular plants) of the Carboniferous period. Her method was to study under the microscope sections of petrified plants; sections she cut with a rather hazardous gas-driven lapidary housed in a hut in the grounds, a safe distance from the College. She illustrated her own papers with great artistic skill, and sometimes colour washed her drawings (Blackwell 1936–1937). Her researches in the early years at Royal Holloway were closely related to those of DH Scott, from whom she received characteristically helpful support and guidance. Scott and Oliver were among the seven men who supported her application for fellowship of the Linnean in 1904–1905. Overall, her research contributed significantly to contemporary understanding of the evolution of the seed-bearing plants from their spore-bearing ancestors, a subject which J Reynolds Green says in his *A History of Botany 1860–1900* was attracting a great deal of attention at the start of the twentieth century, and in need of hard facts to limit some of the wilder speculation that was abroad (Reynolds Green 1909, 123). Margaret's research provided many of those badly needed facts.

Margaret was admitted to the Linnean at the same time as her good friend Ethel Sargant, someone whom Margaret found a constant source of strength and encouragement as she built her department at Royal Holloway. In 1897 the two women spent the summer on a Grand Tour of the leading botanical laboratories of Europe, the high point being a visit to Edward Strasburger in Bonn, where they were able to learn at first hand from the greatest expert on the reproductive anatomy of flowering plants. Observations made in the best European laboratories also helped Margaret decide how she should equip her own teaching laboratory back at Royal Holloway.

With a taste for travel, she made further visits to Europe. She collected from the Middle East and during a year's enforced sick-leave in 1905 she took the opportunity to recuperate in Australia, collecting and bringing back to Royal Holloway's museum a huge number of specimens.

Margaret Benson's illness is significant for another reason. Leave of absence from the College had to be granted at short notice on the advice of her doctors, which threw a huge burden upon two young members of her staff, Helen Fraser and Evelyn Welsford. Fraser had only a few months earlier been appointed by Royal Holloway as Benson's Demonstrator. After Benson's departure for Australia, Fraser was very soon promoted to Assistant Lecturer, understandably finding it a daunting task teaching honours students who had already enjoyed two years' experience of Benson's meticulous training (Izzard 1969, 84).

> *While the students were doing practical work, Miss Welsford was installed in the neighbouring preparation room, surrounded by as many reference works as possible. If a student asked a question, Miss Fraser would suggest that she try to think it out, and then at an appropriate moment, slip unobtrusively into the preparation room, where Evelyn Welsford would be hurriedly looking up the required information.* (Izzard 1969, 86)

Margaret's department was still flourishing when she returned, restored to good health. Fraser and Welsford, who had enjoyed life-changing opportunities during Margaret's absence, remained friends, soon leaving Royal Holloway as they moved to support research and teaching in mycology at Imperial College—that is, until the outbreak of WWI. After war service, their paths then diverged: in post-war years Welsford worked as the government mycologist in Zanzibar, while Fraser remained involved with military and public service in Britain.

Margaret Benson was the best of academics. Unafraid to get her hands dirty, she helped her chosen area of research to make significant advances, and she inspired her students and her young colleagues.

Helen Charlotte Isabella Fraser, Dame Helen Gwynne-Vaughan

Helen's branch of the Fraser clan was closely related to the Scottish nobility. At her worst, and from childhood, she was spoiled, haughty, and manipulative, harbouring throughout her life a scarcely disguised contempt for the mercantile ethos represented by the 'business-classes' (Izzard

1969, 44). The boredom of a childhood and an adolescence spent with governesses was finally relieved when in 1895 this energetic and highly intelligent young woman was allowed to enter Cheltenham Ladies College. At last she was in a more challenging environment which stretched her mind.

Having, at the age of twenty-one, overcome family objections to study mathematics, botany and chemistry, at Kings College, Ladies Department, Kensington Square, London—her family would have preferred her studies to follow a more traditional course in Oxford—she characteristically took on and persuaded the authorities of King's to allow her to join the medical course at its main campus in The Strand. She thereby satisfied her desire to study zoology, which was not taught at the Ladies Department (Izzard 1969, 76–80). She and the one other woman taking the zoology course suffered from not only the continual attentions of the male students but also an absence of any toilet or common room facilities. In her second year, Helen again challenged the King's authorities, this time by winning the Carter Gold Medal, awarded to the best student in botany. But there was a potential hitch before the prize could be awarded because the College authorities had to question their own regulations; the question was, are women eligible? Eventually, it was conceded that they were, and in her final year Helen happily went on to specialise in botany, with subsidiary animal physiology.

In 1903, the year before she graduated, a friend took her to the Botanical Department of the Natural History Museum and introduced her to the Keeper, George Murray. The old man took a shine to Miss Fraser and invited her to lunch, an invitation which she confidently accepted despite the scandalised warnings of her friends, including Annie Lorrain Smith, who worked in the same department and knew Murray well. Helen was not afraid of men and was certainly not a young woman of whom advantage could be taken; the lunch went well and she was given employment as a part time assistant, helping to prepare exhibits for public display. Much more significantly for the direction of her future career, she met a young mycologist, Vernon Blackman, who was one of the Museum's assistant keepers. Alongside her studies at King's she would, when time permitted, help Blackman in an unpaid role in the morning, and then do her paid work in the afternoon. It is perhaps not surprising that her B.Sc. was awarded with only second class honours. By this time, 1904, Blackman was giving an advanced course in mycology at University College and he needed a demonstrator. He chose Helen. When a similar post, but offering greater security, became available with Margaret Benson, Helen moved on in July 1905 to Royal Holloway College. (She was appointed in spite of turning up for interview wearing an elegant brown linen dress and a hat covered in nasturtiums, 'I suppose I looked unacademic', she later admitted.) Helen was well known for displaying the same uncompromising self-confident spirit at meetings of the British Association for the Advancement of Science, wearing bright dresses and large hats when other women were attempting to be unobtrusive.

Helen was now a mycologist. In 1905, with Vernon Blackman as first author, she published 'Fertilization in *Sphaerotheca*' in the *Annals of Botany*. Two more papers with Blackman followed the next year, the second in the *Proceedings of the Royal Society*. Helen was barely 26 years old at the time of her fellowship of the Linnean and, like the 33-year old Emily Berridge (who follows), her promise was at that stage greater than her proven achievement. It was a promise that was however to be fully realised in the next ten years.

Joining the Linnean was consistent with her efforts to ensure there were equal opportunities for women in the sciences. When an opportunity arose, it had to be taken. As befits such a strong minded woman, Helen was outspoken on the matter of women's suffrage, often making speeches and canvassing support. With Millicent Fawcett's niece, Dr Louisa Garrett Anderson, she started a University of London Suffrage Society, a body initially concerned as much with economic and career opportunities as with votes for women, and which, remarkably, was open to both female and male graduates (Izzard 1969, 109). Helen's sympathies were always with Fawcett's suffragists; she found Mrs Pankhurst and her suffragette daughters too emotional and extreme. 'I could have slapped Christabel and Sylvia', she once exclaimed.

In 1907, Helen obtained a D.Sc. from the University of London, but only at the second time of asking, the first piece of work she submitted being judged too lightweight. The D.Sc. was the trigger which prompted her to apply for and obtain a lectureship at University College, Nottingham, then affiliated to London University. She returned to London in 1909 when she took the Head of Department's position at Birkbeck College vacated by the man whom she was to marry two years later, David Gwynne-Vaughan. A palaeobotanist, who numbered D.H. Scott and Marie Stopes among his friends, Gwynne-Vaughan had moved to Belfast, and for most of their brief marriage the couple lived apart, except during university vacations. As war was declared in August 1914, he took the chair at

University College, Reading, and for the first time Helen was able to live continuously with him, commuting daily to fulfil her obligations at Birkbeck, her thoughts turning to the possibility of motherhood. Tragically, their chance to experience normal married life together was brief. By September 1915 David was dead, a victim his family's susceptibility to tuberculosis.

In the earliest days of the war, Helen saw it as her patriotic duty to undertake first-aid training. She passed her exams and joined the Voluntary Aid Detachment (civilians who nursed injured soldiers). In 1915, with the Birkbeck classes depleted of most of their male students, she found time to join bacteriology classes at Kings College Medical School—a humble step for a leading academic. Thanks to her friendship with Louisa and other members of the Garrett Anderson family, the word soon got around that, as her next step, Helen fervently wished to go to France to serve her country. It was not long before this forceful woman, leading female scientist, scion of the Scottish nobility, and soldier's daughter, was invited to be the Controller (France) of Queen Mary's Army Auxiliary Corps. This is not the place to trace Helen's extraordinary story further, except to note that her upward progress continued as she was soon appointed Commandant of the Women's Royal Air Force, 1918–1919, her services to the nation being recognised in 1929 when she was made a Dame of the British Empire. Returning to Birkbeck in 1919 to continue her studies, she was the first women to be awarded the Trail Medal of the Linnean Society in 1920, the citation noting that it was for her research on the morphology and cytology of fungi. In 1921 she was appointed a full professor of the University of London. In spite of being made President of the British Mycological Society in 1928, as a scientist, she is perhaps best remembered her text book, *The Structure and Development of Fungi*, rather than for any discoveries she made. Although she was a meticulous microscopist, she wrongly interpreted the meiotic (nuclear division) process in Ascomycete fungi, a mistake which led many mycologists of her generation along a false trail (Carlile 2005).

Emily Mary Berridge, Quiet and Unassuming

No greater contrast to Fraser could be found than Emily Berridge. Diffident by nature, shy, and needlessly modest about her own achievements, she needed the encouragement of others throughout her life. Her father instilled in his daughters an interest in science. It is said that they enjoyed standing on an insulated stool while he, with his 'electrical machine', made their hair stand on end (Blackwell 1947–1948). A member of the Institute of Actuaries, her father was wealthy enough to pay for Emily to attend Dulwich High School (a Girls' Public Day School, founded in 1878), from where she progressed to Bedford and Royal Holloway Colleges. At the latter she obtained a B.Sc. in physics, while winning the Driver Prize for an essay in botany. It was botany, rather than physics, which she was to pursue for, after two years teaching in York, she returned to London to conduct research under the watchful eyes of Margaret Benson, and also of Ethel Sargant.

From Benson's letters preserved in the archives of Royal Holloway College, the warmth of her support and encouragement of the diffident Berridge is apparent.

> 4 October 1903
>
> *Could you run over and stay the night any day this week and bring the slides and I will see if I can help you to give an account of them.* [Benson continues] *I took the liberty of telling Dr. Scott about it on Saturday and he was very envious.*

And a little later, she wrote:

> 24 February 1904
>
> *Miss Sargant has been down here for two days looking through the Carpinus slides Miss Sanday made.* [Benson continues] *She says experiments ought to be made in staining and she is willing, if you like, to undertake the work, to give you her help.*

Ethel Sargant also invited Miss Berridge to stay with her so that they could study together.[15] The Benson-Berridge relationship has parallels with the Sargant-Robertson one. Benson was acquainted with Berridge's family, clearly having first-hand knowledge of both the younger woman's mother and sisters.[16] Benson's letters similarly maintain a formality, beginning, 'My dear Miss Berridge', never Emily, and ending, 'Yours affectionately, M. Benson'. When Benson was forced to take sick leave from her duties at Royal Holloway College, it was Berridge who was her first choice of

[15] Letter from Ethel Sargant to Emily Berridge, 2 March 1904. Archives of Royal Holloway College.

[16] Letter from Margaret Benson to Emily Berridge, 13 November 1904. Archives of Royal Holloway College.

travelling companion, though Emily declined and Benson's sister eventually accompanied her to Australia and New Zealand.[17]

At one stage Berridge set up a laboratory in her home, very much in the style of Sargant, but by 1906 she realised the cost was too much to bear and she registered as a research student at Imperial College. In 1914 she was awarded a D.Sc. degree.

Berridge's election to the Linnean seems to have been on the basis of insider information for, although she had no publication record at the time (March 1905), two papers were in the pipeline. The first of those appeared the very next month, in the *Annals of Botany* (Berridge 1905). On 16 November of the same year Emily became one of the first women to read a paper before the assembled fellows: her co-authors were Margaret Benson and Elizabeth Sanday; her paper was Part III of a study of the embryology of fossils belonging to the Amentiferae (Part I was a joint effort between Benson and F.W. Oliver). Many other papers were to follow: at one time she set herself the (over)ambitious target of writing one paper every year (Blackwell 1947–1948).

As the shy Berridge nervously contemplated attending her first meeting of the Linnean, Benson consoled her, telling her that she would immediately recognise a friend: '…you will find Dr Scott sitting by the President', then commenting wryly: 'There is no special Probationer's seat'.[18]

Emily Berridge's interests were changed fundamentally by WWI. From January 1916, her focus turned from botany to bacteriology. Working at the University of Liverpool she was part of team studying the agglutination and identification of the bacteria associated with dysentery, a condition all too common in troops invalided home from the Middle East. After the war, she returned to London managing to combine her earlier interests by linking her newly acquired expertise to the pioneering studies of plant pathogenic bacteria then being carried out at Imperial College.

In a long working life Berridge was renowned for her generosity towards, and wise mentoring of, younger female botanists, maybe remembering the help that in her younger days she had been given by Benson and Sargant. She donated microscopes and other small pieces of

[17] Letter from Margaret Benson to Emily Berridge, 'Early November 1905'. Archives of Royal Holloway College.

[18] Undated letter (? March 1905) from Margaret Benson to Emily Berridge. Archives of Royal Holloway College.

equipment for botanical research at Royal Holloway College, and she performed a major service for the botanical community by undertaking for many years the Treasurership of the Botanical Research Fund.

Henderina Victoria Scott, Time-Lapse Photographer

Rina, as she was known to her friends, was born in 1862 into a prosperous South London family. Like Emily Berridge, Rina's interest in botany—and that of her younger sister, Helen Gertrude, for physics—was inspired by a father's fascination with the natural sciences. Their father, Hendericus Klaassen, was the son of a Dutch Pastor living in Hamburg (Anon. 1910). At the age of twenty he had moved to London where, as a corn factor based in London's Corn Exchange, he ran a business which was so successful that by 1874, at the age of forty-six, he was able to retire. Immediately, he began taking courses in chemistry, zoology, and geology at UCL. By 1877 he was elected a Fellow of the Geological Society. Studying the geology of the Croydon area (south of London), with the help of a zoologist, E.T. Newton FRS, he familiarised himself with the fossils of the region's pebbly and sandy beds, a familiarity which in later years must have fuelled many happy hours of discussion with his son-in-law. It is also noteworthy that Hendericus was an ardent supporter of the Croydon Natural History and Microscopical Club, and that he helped to found a school in Croydon for the secondary education of girls, a school associated with the Girls' Public Day School Company.

Helen proved to be an outstanding physicist. In the late 1880s and early 1890s she carried out experiments on electrochemistry under the supervision of the great J.J. Thomson in Cambridge's Cavendish Laboratory. By the late 1890s, she was employed by Newnham College as a demonstrator and lecturer in physics. Among her colleagues in the Cavendish and at Newnham was Philippa Fawcett—daughter of Millicent Fawcett and niece of Elizabeth Garrett Anderson (Gould 1997). The Klaassen sisters' attitude to women's suffrage and allied causes is not known but, whatever it was, Rina's household had, through Helen, a link to major players in the struggle for suffrage, and their politics.

Little is known about Rina's education except that in 1886, aged twenty-four, she was taking classes at the Royal College of Science, South Kensington, and that her teacher in Advanced Botany was Dukinfield Henry Scott. Scott was in the habit of taking his students to the British Museum (Natural History) and to the Jodrell Laboratory at Kew where they could try their hand at research. Rina's first work, concerning the nuclei of the green algae, *Oscillaria* and *Tolypothrix*, was carried out in the Jodrell at this time and seems to have been an integral part of a whirlwind romance, for pupil and teacher were married in 1887. Most likely, the contribution of the novice researcher was small, the paper on the algal studies being published in 1887 under the name of her husband, although Rina's contribution was acknowledged (Scott 1887). In 1898 she published in her own right, with Ethel Sargant, a paper describing their joint investigations of the development of *Arum maculatum* from seed (as opposed to from a tuber).

An accomplished artist, she prepared illustrations for her husband's books, *Introduction to Structural Botany* (1894–1896) and *Studies in Fossil Botany* (1900). Through these works, and the time she spent cataloguing and indexing her husband's collections of fossil slides, she became an expert in palaeobotany, contributing her own research papers on the subject to the *New Phytologist* (1906) and the *Annals of Botany* (1908, 1911).

Rina's greatest distinction was as a pioneer of time-lapse photography, a technique allowing the slowest movements of living organisms to be speeded-up and thereby studied in a way previously impossible. The style of the *Proceedings of the Linnean Society* in Edwardian times was no more, or no less, dry and terse than the journals of contemporary scientific societies. Thus, the *Proceedings* for 1904–1905 sparingly record that, on 16 March 1905, "Mrs DH Scott, F.L.S., exhibited animated photographs of plants taken by the Kammatograph, showing the natural movements of plants accelerated so as to be readily followed by the eye" (Anon. 1904–1905). Among the nine 'animated photographs' shown were: stamens of *Sparmannia africana* responding to light; leaves of *Mimosa pudica* responding to touch; buds of *Fuchsia* opening; circumnutation of the stem and petioles of *Maurandia* (that is, twining around a stick); and a humble-bee fertilising a *Scabious* flower. Maintaining its objective tone, the *Proceedings* noted that, 'A discussion followed in which Dr Scott, Rev TRR Stebbing, Mr EM Holmes, Mr J Hopkinson and the President [Prof WA Herdmann FRS] took part, Mrs Scott answering questions.'

What the journal failed to convey was that the fellows of the Linnean had witnessed something extraordinary, a huge step forward in the history not only of biology but also of cinematography. In Rina's own words, 'After a warm rain we often say that we can almost see our plants growing; by means of this adaptation of the cinematograph we literally can' (Scott,

Mrs. 1907). The fellows had witnessed time-lapse films that were longer and of a quality far higher than had ever been previously seen.

There had been earlier, cruder attempts at time-lapse photography. In the 1870s Eadweard Muybridge had entranced audiences with his very short, and repeated, sequences of a galloping horse (proving for the first time that all four hooves can be off the ground at the same time). Members of the Linnean Society had been able to wonder at such tecniques when, in June, 1889, Muybridge had demonstrated to one of its meetings 'Animal Locomotion', a talk illustrated by his projections 'of instantaneous photographs…to which motion was imparted by means of a zoopraxiscope' (Gage and Stearn 1988, 86). The German plant physiologist, Wilhelm Pfeffer, had in 1898 made four short, grainy films of plant movements, including the growth and wilting of tulip flowers, and the development of the root system of broad bean seedlings (*Vicia faba*) over the course of 11 days.[19] While Muybridge used several cameras, each incorporating a large and unwieldy glass plate, Pfeffer used a standard cinematograph and celluloid film. Although he was a gifted technician, Pfeffer farmed out much of the technical development of his cinematograph (Gaycken 2012).

Possibly inspired by Pfeffer (an ex-student of von Sachs', as was her husband, Dukinfield), Rina began her experiments in 1902 using celluloid film but quickly realised that it did not survive well in the moist atmosphere of the glasshouse in which her plants were growing. The key to Rina's success was her decision to use instead a recently patented device (1898) called a 'Kammatograph' (Fig. 5.4).

The Kammatograph combined a camera with a projector. A spiral row of images was first recorded on a circular glass disc. After the images were developed and fixed they could be projected. Rina's Kammatograph could take 354 photographs per disc, giving about 30 seconds of moving pictures (for most subjects, she took one photograph every 15 minutes). As she admitted, it was very laborious and time consuming; 'I have only about a dozen successful plates as a result of over three years labour' (Scott, Mrs. 1907). Most importantly, her subjects were plants grown under natural conditions in a glasshouse.

Rina's hard and innovative work paid off. After more than a year's experimentation she was able in 1903 to publish her results in the *Annals*

[19] Pfeffer often taught his students using filmless 'projections' of living organisms, such as swarming zoospores.

Fig. 5.4 Kammatograph. The device was invented and patented by Leonard Kamm of Powell Street, London. (Source: From Jones, Claire. 2010. Bodies of Controversy. Women and the Royal Society. *HerStoria Magazine*, 6: 20–24)

of Botany (Scott 1903). In June 1904, she displayed her work to an audience at an evening Conversazione of the Royal Society, which amounted to the annual 'Ladies Night'. On 19th August, she exhibited her work at the Cambridge meeting of the British Association for the Advancement of Science and, completing a busy year, on 2nd December she gave the same programme at the Holmesdale Natural History Club (almost certainly invited by her old friend Ethel Sargant).[20] In March 1905, and newly

[20] https://wfpp.cdrs.columbia.edu/pioneer/ccp-henderina-victoria-scott. Holmesdale Natural History Club. *Proceedings of the Holmesdale Natural History Club for the Years 1902, 1903, 1904, 1905*. Reigate: Reigate Press, 1906.

elected a fellow, she was able to show her films at the Linnean Society (Jones 2010).

The latter was not the only occasion at which the Linnean was graced with her demonstrations of plant movements for in an account of a reception held at Burlington House in 1907 to celebrate the bicentenary of Linnaeus' birth, the *Proceedings* record that, in addition to displays of Linnaeus' artefacts, Mrs DH Scott FLS showed in the lobby her, 'Animated photographs of plant-life shown by the Kammatograph' (Anon. 1906–1907). Such early time-lapse films were not only entertaining but, as Pfeffer observed, they had a 'high didactic value'. Soon, time-lapse photography would also be an important research tool.

The men who signed Rina's Certificate of Recommendation for fellowship were F.W. Oliver and Arthur Lister (both family friends), Albert Charles Seward (a palaeobotanist), and Francis Darwin, a keen student of photography, and someone exceptionally well placed to appreciate the difficulties Rina had had to overcome, and the advances she had made. Francis had jointly authored with his father a book, *The Power of Movement in Plants*, in the preparation of which he had studied carefully Pfeffer's experiments on plant movements (if not Pfeffer's photography) (Ayres 2008, 122). Francis' brother-in-law, WC Crofts, was a leading British photographer and in the early summer of 1889 Francis had visited, in Paris, Etienne-Jules Marey, an animal physiologist who was for the first time putting multiple images onto film strips' (Lefebvre et al. 1999).

Unfortunately, 'Rina's status was predicated not on her science but on her identity as wife of an eminent scientist and hostess of scientific gatherings; as a result, her status as a scientific peer was easily disrupted and rendered secondary' (Jones 2016). She was all-too-easily hidden by her husband's large shadow, her original contributions to the history of botany overlooked.

A Coherent Group?

There emerges from this chapter a group of female botanists who certainly knew each other in twos, threes, or possibly more, *before* their fellowships of the Linnean. Ethel Sargant connected them and, together with influential male botanists such as DH Scott and FW Oliver, whose support of women botanists is the subject of Chap. 8, she nurtured them. Whatever their age and experience—and Margaret Benson was four years older than

Ethel—the women looked to Ethel for guidance in their careers, for constructive criticism of their research, and, above all, for friendship.

At the annual meeting of the BAAS in 1913, Ethel Sargant was given the great honour of being chosen to be the first female President of the Botanical Section (Section K): in the photo taken at the meeting (Fig. 5.5) are seen not only male members of the botanical 'Establishment' of the day, such as Scott and Oliver, but also many of the female botanists whose careers were influenced by Sargant. It is entirely appropriate that she sits at the front and centre.

On the 16th January, 1918, Ethel Sargant died in Sidmouth, Devon, a small seaside town she had often visited when needing to restore her health and spirits. This time the sea air failed her. Her brother, Walter Lee Sargant, Headmaster of Oakham School, was with her at the end. There was shock and sadness within the small botanical community which she had done so much to foster. D.H. Scott wrote to Gulielma Lister, 'My wife had letters from Mrs EB [Edmund Beale] Sargant and Miss [Emma] Turner and I heard of it from Dr Arber whose wife has been in constant communication with Sidmouth'. …[I had the] greatest admiration for her abilities and character and we had been friends since she first came to Kew at the end of 1892'.[21] There was shock and sadness also in the wider scientific community, where Ethel was well known. During the war she had worked for the Board of Trade, compiling lists of women graduates and their skills, matching them with war work for which they were particularly fitted. At the time of her death Ethel was serving as President of the British Federation of University Women, founded by the in 1907 by the Manchester University chemist, Ida Smedley.

Ethel Sargant was so much more than a simple botanist, she was a mother figure.

[21] DH Scott to Miss Gulielma Lister, 24 January 1918, 'Correspondence of Annie Lorraine Smith' Archives of the Natural History Museum.

Fig. 5.5 Ethel Sargant, centre front, among botanists at the 1913 meeting of the British Association for the Advancement of Science. Also in the front row are (left to right); G.S. West, R.H. Yapp, O. Stapf, J. Reinke, D.H. Scott, and F.W. Oliver. (Source: Courtesy of the Hunt Institute for Botanical Documentation, Carnegie Mellon University, Pittsburgh, PA)

References

Anonymous. 1903. *Essex Naturalist* 13: 182.
———. 1904–1905. *Proceedings of the Linnean Society* 117: 10–11.
———. 1906–1907. *Proceedings of the Linnean Society* 119: 63.
———. 1910. Obituary. Hendericus M Klaassen FGS. *Geological Magazine* 191.
Arber, Agnes. 1919. Obituary. Ethel Sargant. *New Phytologist* 18: 120–128.
———. 1927. Ethel Sargant. *Girton Review, Michaelmas* 1927: 17–26.
Ayres, P.G. 2008. *The Aliveness of Plants. The Darwins at the Dawn of Plant Science.* London: Pickering & Chatto.
Berridge, Emily. 1905. Two New Species of *Spencerites insignis. Annals of Botany* 19: 273–280.
Blackwell, Elsie M. 1936–1937. Obituary. Margaret Jane Benson. *Proceedings of the Linnean Society of London* 149: 186–189.
———. 1947–1948. Obituary. Emily M Berridge. *Proceedings of the Linnean Society* 160: 68–70.
Burek, Cynthia V. 2009. The First Female Fellows and the Status of Women in the Geological Society of London. *Geological Society, Special Publications* 317: 373–407.
Carlile, M.J. 2005. Two Influential Mycologists: Helen Gwynne Vaughan (1879–1967) and Lilian Hawker (1908–1991). *The Mycologist* 19: 129–130.
Crawford, Elizabeth. 1999. *The Women's Suffrage Movement: A Reference Guide 1866–1928.* London: UCL Press.
Forrester, J., and Laura Cameron. 2017. *Freud in Cambridge*. Cambridge: University Press.
Gage, A.T., and W.T. Stearn. 1988. *A Bicentenary History of the Linnean Society of London.* London: Academic Press.
Gaycken, O. 2012. The Secret Life of Plants. Visualising Vegetative Movement, 1880–1903. *Early Popular Visual Culture* 10: 51–69.
Gould, Paula. 1997. Women and Culture of University Physics in Late Nineteenth-Century Cambridge. *British Journal for the History of Science* 30: 127–149.
Izzard, Molly. 1969. *A Heroine in Her Time: A Life of Dame Helen Gwynne-Vaughan 1879–1967.* London: Macmillan.
Jones, Claire. 2010. Henderina (Rina) Scott. Botanist and filmmaker. *HerStoria* Summer: 40–43.
Jones, Claire G. 2016. The Tensions of Homemade Science in the Work of Henderina Scott and Hertha Ayrton. In *Domesticity in the Making of Modern Science*, ed. Donald L. Opitz, Staffan Bergwik, and Brigitte Van Tiggelen, 84–104. London: Springer.
Lefebvre, T., J. Malthete, and L. Mannoni. 1999. *Letters from Étienne-Jules Marey to Georges Demenÿ 1880–1894.* Paris: AHFRC.

Rayner-Canham, Marelene F., and G.W. Rayner-Canham. 2003. Pounding on the Doors: The Fight for Acceptance of British Women Chemists. *Bulletin of the History of Chemistry* 28: 110–119.

Reynolds Green, J. 1909. *A History of Botany 1860–1900*. Oxford: Clarendon Press.

Robinson, S., and Kate Perry. 2008. Science at Girton – Ethel Sargant. *Girton College Annual Review*: 21–23.

Sargant, Ethel. 1900. Women and Original Research. *Frances Mary Buss School's Jubilee Magazine*.

———. 1901. The Inheritance of a University. *Girton Review*: 9–21.

Schmid, Rudolf. 2001. Agnes Arber, née Robertson (1879–1960): Fragments of Her Life, Including Her Place in Biology and in Women's Studies. *Annals of Botany* 88: 1105–1128.

Scott, D.H. 1887. On Nuclei in *Oscillaria* and *Tolypothrix*. *Journal of the Linnean Society, Botany* 24: 188–192.

Scott, Rina. 1903. On the Movements of the Flowers of *Sparmannia africana*, and Their Demonstration by Means of Kinematograph. *Annals of Botany* 17: 761–777.

Scott, Mrs Dukinfield H. 1907. Animated Photographs of Plants. *Journal of the Royal Horticultural Society* 32: 48–51.

Stearn, W.R. 1960. Mrs Agnes Arber: Botanist and Philosopher, 1879–1960. *Taxon* 9: 161–263.

Swanwick, Helena M. 1935. *I Have Been Young*. London: Victor Gollancz.

CHAPTER 6

Approved by Mrs Farquharson?

Marian Farquharson argued for the right of 'duly qualified' women to be able to join scientific and learned societies. If she had been called upon to cite a woman who was duly qualified, she could have done no better than to point to Maria Ogilvie-Gordon, born and raised in a part of Aberdeenshire Marian lovingly adopted as her home after marriage. Maria was by 1904–1905 outstandingly qualified to be accepted into scientific society, having already accumulated three degrees and published more than nineteen papers. Similarly well qualified, by Mrs Farquharson's criteria, was Lilian Veley, an Oxford graduate who subsequently obtained a D.Sc. degree from Dublin. Both Maria and Lilian were daughters of the clergy, as was Marian, surely one further recommendation in Marian's eyes. Alice Embleton would, however, not have been approved of. Although Alice's intellect might have earned some admiration, for she had obtained a degree from the University College of South Wales (Cardiff), and she deserved some Christian sympathy for the financial hardships she suffered, Marian would surely not have approved of the company Alice kept, which was with lesbian suffragettes.

P. Ayres, *Women and the Natural Sciences in Edwardian Britain*,
Palgrave Studies in the History of Science and Technology,
https://doi.org/10.1007/978-3-030-46600-8_6

Maria Matilda Ogilvie-Gordon, Dame, Alpine Geologist

Maria Ogilvie was born, in 1864, in the tiny hamlet of Monymusk, barely nine miles from the Farquharson estates at Alford (Chap. 4). Both Maria and Marian were the eldest daughters of clergymen who had been blessed with large families, clergymen who brought up their daughters to be strong-minded and to fight for what they believed in, which in both cases included women's rights. Both flirted in early adulthood with the study of music. Although their periods of residence in Aberdeenshire hardly overlapped, the two women certainly knew each other for Maria was the first secretary of the Scottish Association for the Promotion of Women's Public Works, which Marian founded in 1902. Whether they were related by marriage through the Ogilvie branch of Robert Farquharson's family is uncertain. The similarities end there, however, for while Marian had no formal education, Maria was sent to a boarding school in Edinburgh at the age of only nine years.

Participation in education was a defining characteristic of Maria's family; her father, the Rev Alexander Ogilvie, and also five uncles, were heads of schools or colleges. After 18 years as the schoolmaster at Monymusk (1854–1872), Alexander was appointed Head of Robert Gordon's Hospital [now College] in Aberdeen; his family, including the eight year-old Maria, moved to the city; by 1888 his fortunes were sufficient to build a large country residence, 'Darroch Learg', at Ballater (MacPherson 1895, 290). Maria's oldest brother, Francis Grant, was the first Principal of Heriot-Watt College in Edinburgh, and became a Fellow of the Geological Society.[1]

After leaving her school in Edinburgh, Maria's immediate destination was the Royal Academy of Music in London; she was only eighteen and soon forsook her piano studies to return to Scotland and enter the world of the natural sciences (Burek 2009). By studying at Heriot-Watt, where her approved courses included physics, chemistry, and biology, she was able to matriculate, that is register, for a B.Sc. degree from the University of London. Completing her undergraduate work in London, at University College, she studied geology, botany, and zoology, one of her teachers of

[1] Heriot-Watt College dates from 1855, though its origins lie in the Edinburgh Mechanics Institute started in 1821.

the last subject being the influential Edwin Ray Lankester FLS. Maria graduated in 1890.

It was Maria's ambition to broaden her experience by studying abroad. Her chosen country was Germany and in 1891, supported by a grant from the Royal Society, she attempted to enter the University of Berlin (Wachtler and Burek 2007). Her approaches failed, however, despite the intercession of one of the University's most distinguished scientists, Professor Baron Ferdinand von Richthofen, an admirer of the young Scottish woman's intellect and tenacity. Guided by the Baron, Maria next tried the University of Munich, where she was marginally more successful, being allowed to carry out private research, but not to register formally in the University. In July 1891, the Baron and his wife took Maria on a five-week field trip to the Dolomites, a mountainous area in the northern extreme of modern Italy with which he had become familiar some thirty years earlier when his own extensive field work had led him to the conclusion that the Dolomites were essentially fossilized coral reefs. Encouraging Maria to concentrate her efforts on geology, rather than zoology, the Baron introduced her to the guides, innkeepers, and local fossil collectors who were to prove so important when he and his wife finally left her on her own after they returned to Berlin.

Maria's work was physically demanding, often involving pre-dawn starts, a day's work in a lonely and dangerous situation, followed by a long trek back to her overnight resting place, usually a simple inn—all the while carrying the heavy specimens she had collected. The local roads gave only poor access to the mountains. In spite of such difficulties, Maria returned to the region in the summers of 1892 and 1893 to continue her researches. Carefully planning her programme, she learned German and took lessons in rock climbing from a well-known local expert, Josef Kostner, who often accompanied her on the most hazardous expeditions. By late 1893 she was able to submit to her *alma mater*, the University of London, a thesis, 'A geological survey of the Dolomite region of the Tyrol', on the strength of which she became the first woman to be awarded a D.Sc. degree by the University. Based on well-tried geological methods, such as the analysis of fossil deposits, her Ph. D. and also much of her later work drew upon the newly emerging theories of plate tectonics in order to explain folds in the wider Tyrolean landscape.

In 1895 she married John Gordon, an Aberdeen physician fourteen years older than herself, and they raised three children in that city. Soon however the children were judged old enough to accompany Maria and her husband on summer holidays in the Tyrol; holidays which of course contained a large element of geologising. There was clear evidence of the scientific progress she was making when, in 1897, with the support of both her old teacher, Lankester, and Sir Archibald Geikie, one of her papers was published in the prestigious *Philosophical Transactions of the Royal Society* (Ogilvie-Gordon 1897). It was a proud moment for Maria, and another one followed in 1900 when, at long last, the University of Munich agreed to admit women, and Maria became the first to whom it awarded a Ph.D.

Although her field work was interrupted by WWI, she returned to the Dolomites in the 1920s. In the course of her long career, she published more than thirty original research papers, wrote both academic and popular books about geology, and translated standard German texts into English. In 1919 she was among the first women to be admitted to the Geological Society, and in 1932 that Society recognised her outstanding work by awarding her its Lyell Medal.

In addition to the demands of geology and her family, Maria dedicated much of her time to working for women's causes, no doubt motivated by the sexual discrimination she had experienced as a young woman. Her particular interests were the education of girls and the protection of women in work, as, for example, their rights during pregnancy. Hygiene and safety, for both sexes, were among her concerns, as too was the importance of tranquillity in a child's home environment. In 1908 she published a *Handbook of employment specially prepared for the use of boys and girls on entering the trades, industries and professions* (Aberdeen: Rosemount Press), in which she argued for the establishment of employment bureaux to advise school leavers on courses available for the continuation of their technical or commercial education.

Apart from the obvious recommendation of her exceptional qualifications, her reputation for hard work and meticulous attention to detail made her a natural candidate for high office. Maria was a vice-president of the International Council of Women in 1904, and, in 1916, she was President of the National Council of Women of Great Britain and Ireland. Immediately following the end of WWI, she was heavily involved, on behalf of the Council for the Representation of Women, in negotiations at the newly formed League of Nations. For these and many other activities, King George V made her a Dame in 1935.

Cynthia Burek, who has made a close study of Maria's life, remarked tellingly that she, '…was of independent means and therefore had the ability to carry on her work with the help of servants and family members' (Burek 2009). Many of her contemporaries were equally blessed with good fortune and a good brain but they have been forgotten. Maria Ogilvie-Gordon stands out because through hard work and dedication she made the very best use of her talents and the opportunities available to her.

Lilian Jane Veley, Steamboat Lady

Lilian was the youngest of the twelve children of the Rev. John Nutcombe Gould, the wealthy rector of Stokeinteignhead, Devon, and his wife, Katherine, the daughter of a major-general.[2] Educated at home, where for a period a German governess was employed, it was not until the advanced age of twenty-nine, in 1890, that Lilian won a scholarship, worth £50 over two years, and entered Somerville Hall [now College], Oxford. She was the first 'Adult Student' to come to Oxford following efforts made to recruit the brightest of these by engaging them in extra mural studies. Lilian had come to the attention of a young Edward Bagnall Poulton when, in 1889, he was teaching an extension course at Newton Abbott, Devon, close to Lilian's home which since the age of twelve had been at Lustleigh in the parish of Bovey Tracey. Poulton's subject was 'Colours in Animals', similar to the title of a book he was about to publish (Poulton 1890).[3] He quickly noticed Lilian's 'singular powers of accurate and original observation', and encouraged her to apply for the scholarship at Somerville.[4]

In 1894 she obtained the equivalent of a First Class degree in Natural Sciences ('equivalent' because women were not allowed to graduate formally), with a specialization in animal morphology. Clearly an outstanding student, she wrote her first two papers while still at Oxford. The first shows the influence of Poulton for it was on colour in the larvae of lepidoptera, while the other was on an amoeba, *Pelomyxa palustris*. Not only that, Somerville's *Report for 1891* records that

> Miss Lilian Gould, while still in residence, was invited in October to read a paper before the Entomological Society on observations on Lepidoptera

[2] www.devonhistorysociety.org.uk/veley-mrs-lilian/.

[3] Poulton was in 1893 appointed Hope Professor of Zoology in the University of Oxford. He was President of the Linnean Society from 1912 to 1916.

[4] Archives of Somerville College, Oxford, SC/AO/RG/SF/RI-025.

> made at the suggestion of Mr. Poulton. Miss Gould is, we believe, only the second lady who has received this honour.

Oddly, when Lilian's Certificate of Recommendation for a fellowship of the Linnean Society was completed, Poulton's name was not among those who signed; however, E Ray Lankester's was. The second is more easily explained than the first.

Lankester could have known Lilian in any one of several contexts. He was the Linacre Professor of Comparative Anatomy at Oxford during Lilian's time at Somerville, so he may have known her in that context. Furthermore, not only were they both elected to the Ashmolean Society on the same day, 19 February 1891, but in the last year of Lankester's residence in Bradmore Road, north Oxford (1891–1896), living almost directly across the road from him was the newly wed Mrs Veley (née Gould). To add to the fabled coziness of that part of Oxford, the end of Lankester's garden abutted directly on that of the Poultons, whose house fronted Banbury Road. The families were such good friends that Poulton 'made rock-work steps to help in getting over the party-wall dividing the ends of our gardens, and often used to take the children over to see my friend' (Poulton 1919, 39).

Running through the list of fellows who signed Lilian's Certificate there are others with Oxford connections. And connecting those names—Druce, Evans, Vines, and Raphael Weldon—was The Ashmolean Natural History Society of Oxfordshire, an organisation, similar to those described in Chap. 2, which brought together men *and* women who wished to share their enthusiasm for the natural world, whether it was through lectures, expeditions, access to a library, or simply through socialising. All the men, save possibly Evans, were members; Lilian was not only the Ashmolean's efficient and reforming Secretary from 1895 to 1905, but on retiring from that post she was invited to be its President for 1906–1907 (Bellamy 1908, 104).

Apart from the many onerous administrative responsibilities which fell on her shoulders, she often gave lectures, including special 'Christmas Lectures for Children' which drew audiences of 50–100. While E.B. Poulton had been the Society's President from 1887 to 1891, predating Lilian's Secretaryship, the man who had done more for the Ashmolean Society than anyone else was the Oxford pharmacist, George Claridge Druce. A founder of both that society and of the Botanical Exchange Club (the forerunner of the Botanical Society of the British

Isles), of which he was Secretary from 1903 to 1932, Druce put extraordinary effort into, and took the greatest delight from, his personal connections (Allen 1994). He knew everyone in British botany, both amateurs like himself, who were the mainstay of the Exchange Club, and professionals such Sydney Vines, Oxford's Sherardian Professor of Botany from 1888 to 1919. As the mayor of the city of Oxford in 1901, Druce used his position to celebrate and promote the Ashmolean Society, all the while extending his own influence (Marner 2002).

In 1895 Lilian had married a future Librarian and Deputy-President of the Ashmolean Society, Victor Herbert Veley, a physical chemist whom she had met while a student in Oxford. Veley took her away from entomology and introduced her to microbiology, by way of 'faulty rum'. Batches of West Indian rum were frequently spoiled, at great cost to the producers, but the reason was unknown until the work of the Veleys. They showed that the cause was not faulty casks, as had been suspected, but was instead a contaminating bacterium named by the Veleys as *Coleothrix methystes*, which they were able to culture *in vitro*. It could live in a 75% alcohol solution, such as rum, because it was surrounded by a protective gelatinous coat. Papers appeared in *Nature*, and the couple wrote a short book, *The Microorganisms in Faulty Rum* (1898), the latter bearing only Victor's name on the cover but both Victor and Lilian's name on the Frontispiece (Veley and Veley 1898).[5] In gratitude for advice the two novice microbiologists had received while completing their work, the book is dedicated to one of the most distinguished microbiologists of the day, Emil Christian Hansen, Director of the Copenhagen's Carlsberg Institute. As explained in the Veleys' Preface, Hansen had examined their 'materials, cultures and preparations', confirming their results. After the Veleys left Oxford for life in London, they continued, with ever growing confidence, to research microbiological problems associated with food and drink, a steady supply of such problems being provided via Victor's directorship of the Baddow Brewery Company of Chelmsford, Essex.

In 1905, Lilian became a 'steamboat lady', thereby finally acquiring a formal degree, a D.Sc. from Trinity College, Dublin. Like the chemist Gertrude Elles, she was just one of many frustrated female 'graduates' of Oxford and Cambridge in this period who took Dublin degrees '*ad*

[5] Also note, 1897, A bacterium in rum, *Society of the Chemical Industry Journal*, 16, 626; 1897, A bacterium in strong spirit, *Nature* 56, 197; and 1900, The micro-organism in faulty rum, *Nature* 61, 468–469.

eundem'; they were simply required to apply to Trinity, take a ferry (steamboat) across the Irish Sea, and then spend one or more nights in the city.

The origins of Lilian's interest in women's suffrage are unknown, and may have developed long after her fellowship of the Linnean, but Elizabeth Crawford's *Reference Guide* on the subject of suffrage notes that in 1913 Lilian belonged to London's Halcyon Club, a club for professional women from the worlds of science, medicine, and the arts. Lilian was at that time a member of the committee of both the Marylebone and Paddington branches of both the Conservative and Unionist Women's Franchise Association (CUWFA) and the Church League for Women's Suffrage (Crawford 1999, 121). The political affiliations or sympathies of Linnaeus' Ladies, where they are known, tend to be Liberal, so Lilian's membership of a Conservative organisation sets her apart. Through her association with the CUFWA she was drawn into work for the Red Cross during WWI, soon becoming a Commandant of the British Red Cross.

An ability to observe the natural world accurately and in detail is shared by a majority of successful biologists. Lilian, it seems, had an eye for birds. Indeed, the first reference to her in adult life concerns birds and is found in her father's parish magazine where in 1888 she was campaigning for the protection of owls.[6] Her interest in avian matters continued. During married life in Oxford, she wrote to the *Oxford Times* (1900) protesting about the use of birds' feathers in ladies' hats, and sent letters to *Nature* (1902) relating observations of 'Birds capturing butterflies in flight' and 'Birds attacking butterflies and moths' (Veley 1902a, b). In 1909, the 'Correspondence' section of *Country Life* magazine contained a picture of her (and an unknown man) ringing 'young wheatears' on Palling Marshes (Anon. 1909, 814). The photograph had been submitted by Emma Turner, a distinguished bird photographer who is one of the subjects of Chap. 9—whether the women knew each other before, or as a result of, being made Fellows of the Linnean Society in 1904–1905 is unknown.

Among Lilian's many talents, and dependent on keen powers of observation, she was an exceptionally able photographer. She was an advisor to the photographers' section of the Lyceum Club (formed in 1908) and exhibited her pictures at the Royal Photographic Society's Annual Exhibitions, although unlike Emma Turner, whose work is discussed in Chap. 9, she was never awarded the Society's Gold Medal. Lilian's subjects were posed, typically static and far removed from the dynamic

[6] www.devonhistorysociety.org.uk/veley-mrs-lilian/.

inhabitants of the natural world captured by Turner. The Society's Catalogue tells that in 1908 she displayed pictures entitled, 'Mozambique Guenon' (an east African monkey), and 'Head of french bulldog', while in 1909 she showed 'Head of wire-haired terrier'; 'Swan in evening light'; 'Jimmy' (a bulldog); and 'Studies of Siamese cat'.

The last subject is significant because Lilian is popularly remembered today not as a microbiologist but as the woman who introduced Siamese cats to Britain. This happened in 1884 when a breeding pair was given to Lilian by one of her brothers, Edward Blencowe Gould, Consul-General in Bangkok (Siam = Thailand). The striking appearance of the cats won a fast growing band of devoted owners so that by 1901 Lilian and others were able to found the Siamese Cat Club. All of which tends to detract from her very real achievements as a pioneer of microbiology.

Alice (Alick) Laura Embleton, a Poor Outsider

The first twenty five female fellows of the Linnean Society typically had few financial worries. The one who was clearly an exception, in that her career was seriously compromised by a lack of money, is Alice Embleton. On 14 October 1917, in a letter marked 'Private', Alice confessed her troubles to Benjamin Daydon Jackson, ex-Botanical Secretary of the Linnean Society, from whom she had received an 'agitating' letter pointing out that her annual contribution to the Society was £9 in arrears. Alice replied:

> *I hate owing it like this and was always hoping to be able to keep it up. ... I must resign now. Frankly I cannot afford it and there is no disguising the fact.*[7]

The Council of the Linnean sympathetically cancelled Alice's arrears. However, on 31 October 1917 another letter followed from her. While profusely thanking Jackson, she confirmed her decision, 'I am so sorry to have had to withdraw from the Linnaean [*sic*]'. Alice needed to secure some income from employment, but it seems that for several years she had been unable to work. A likely explanation is that, about this time, she was fully occupied nursing her father, who died in 1917.

It was not the first time that Alice's life had foundered for a lack of money. At the age of fifteen, and in what must have involved a

[7] Archives of the Linnean Society.

heartbreaking parting, 'pecuniary reasons' forced her to leave Sutton High School, a Girls' Public Day School (founded 1884).[8]

Alice was however a young woman of exceptional resolve, as well as exceptional talent. Somehow, she won scholarships enabling her to leave her four brothers and sisters at home in Ewell, Surrey, where her father was a relatively poorly paid stationery engraver, to study from 1895 to 1899 at the University College of South Wales and Monmouthshire, the forerunner of Cardiff University. She was taking advantage of a college which, from its charter and opening in 1883, was second only to University College, London, in offering equal opportunities to both sexes. The college's first intake of students comprised 109 men and 49 women; its first hall of residence for women opened in 1885; and in 1904 it appointed the first female professor in Britain, Millicent McKenzie.[9] On Alice's course there were 5 women and 16 men.

Alice graduated with a B.Sc. first class, on the strength of which she won a Research Fellowship from the Royal Commission for the Exhibition of 1851—she was the only woman to win such an award that year (Anon. 1900). The fellowships, worth £150 per annum, were normally tenable for two years although the Commission's records indicate that she was a fellow from 1900 to 1903. Perhaps she was slow to take up the fellowship. Papers she published in the *Journal* of the Linnean Society and in its *Transactions* (Embleton 1900, 1901) show that at least in the Spring of 1900 she was working in the laboratory of Professor Thomas George Bond Howes at the Royal College of Science.[10] In the years preceding the granting of fellowships to women both he and David Sharp, from Cambridge's Museum of Zoology, kindly read Alice's entomological papers to meetings of the Linnean. When Howes died in 1905 she lost a key supporter.

The Royal Commission's fellowship enabled Alice to move to the Balfour Laboratory in Cambridge to study 'The parasites of Coccidae [Scale Insects] and other researches in economic entomology'. Two of her four sponsors for admission to the Linnean in 1904–1905 were entomologists; David Sharp and the unattached millionaire Nathaniel Rothschild—the latter's great family wealth freed him to develop expertise in entomology, and botany, and to become an early and leading

[8] https://scolarcardiff.wordpress.com/2018/03/05/scientist-and-suffragist/.

[9] McKenzie was the first professor appointed by a fully chartered UK university. Her subject was education.

[10] Howes was a member of the Essex Feld Club.

conservationist (Hobhouse 2002, 268). In 1904, Alice won another prestigious award, a Mackinnon studentship of the Royal Society, '… for research on parasitism and hyperparasitism of Aphidae and Coccidae' (Anon. 1904). To win the award was judged an unique distinction for a woman; there is also a reference at about this time to Alice being a sub-editor of the *Zoological Record*.[11]

For unexplained reasons, Alice's career then took a sharp change in direction as she became a researcher in the Cancer Research Laboratories of the University of Liverpool. The outcome of her time in that city was two papers (1905, 1906) published in the highly regarded *Proceedings of the Royal Society*, the first having been communicated by John Bretland Farmer FRS, the second by one of the Royal's most distinguished fellows, Liverpool's Holt Professor of Physiology and future Nobel Prize winner, Charles Scott Sherrington (Moore and Embleton 1906; Walker and Embleton 1906). A further paper, in 1908, was published in volume one of the *Quarterly Journal of Experimental Physiology*, and was from the Laboratory of Cytology (Walker and Embleton 1908).

Alice's career was flourishing by any standards, her connections impeccable, yet only one year later her mysterious descent into oblivion had begun. She was writing to Daydon Jackson, this time asking the Council to publish her paper on 'Spermatogenesis in rat', the work being condensed from two she had submitted a long time before to the University of London in fulfilment of the requirements for the award of a D.Sc. degree, her letter bearing an address in South Kensington and a sense of desperation. By the census of 1911, when her occupation was still listed as 'cancer researcher', her address is recorded as Barnsley, Yorkshire (where she had family connections) (Fig. 6.1). The census records that she was a 'visitor' in the home of Alderman Charles Wray and his architect daughter, Celia. The resourcefulness, or luck, of this resourceful and talented women had finally been exhausted.

One of the few facts known about Alice at this time is that both Celia Wray and Alick J Embleton [*sic*] were signatories to the constitution of the all-woman Foosack League when it was established in 1910. Membership was open only to suffragists and, by inference (since it had to be kept secret), lesbians (Hamer 2016, 56). Alice had other links to the radical politics of the day since her sister, Florence, was secretary to Rosalind

[11] Archives and website of Sutton High School, Surrey.

Fig. 6.1 Alice Embleton, on the left, and Celia Wray, in the centre, of a group of suffrage supporters photographed in Barnsley, Yorkshire, after the General Election of 1910. Sir Joseph Walton, the successful Liberal candidate, had voted in favour of the Women's Enfranchisement Bill of 1908. In the election, 7560 signatures were separately collected in favour of women's suffrage. (Source: https://www.flickr.com/photos/lselibrary/31268307763)

Howard, Lady Carlisle—the so called 'Radical Countess'.[12] Rosalind Howard was the daughter of Henrietta Stanley (Baroness Stanley of Alderley), a 'lady visitor' at Queen's College, and a founder of Girton College, along with Barbara Bodichon and Emily Davies. The Radical Countess was President of the Women's Liberal Federation from 1894 to 1902 and 1906 to 1915. For Alice, one further link to the suffrage movement was via a cross-dressing actress, Vera (Jack) Holme, a close friend of both herself and Celia, and sometime chauffeur to the suffragette leader, Emmeline Pankhurst.

Alice's trail goes cold from 1911 until one final event, her death in 1960, aged 83, at Bradfield, Essex. Much of Alice's, or Alick's, life therefore remains unknown, but among the first female fellows she was clearly one of the most talented, and unorthodox. Whether her sexual orientation was so overt as to be an impediment to a career must remain an unanswered question. Mrs Farquharson, and possibly also the conservative Lilian Veley, would not have approved of Alice's lifestyle.

Among the twenty-five there may have been women who cloaked their sexuality to the extent that it did not hinder their careers. But that remains their secret.

References

Allen, D.E. 1994. Oxford and Druce. *Botanical Society of the British Isles News* 67: 41–45.

Anonymous. 1900. Notes. *Nature* 62: 380.

———. 1904. *Yearbook of the Royal Society*. London: Harrison and Sons.

———. 1909. *Country Life*, May 29, p. 814.

Bellamy, F.A. 1908. *A Historical Account of the Ashmolean Natural History Society of Oxfordshire, 1880–1905*. Oxford: Published by the Author.

Burek, Cynthia V. 2009. The First Female Fellows and the Status of Women in the Geological Society of London. *Geological Society Special Publications* 317: 373–407.

Crawford, Elizabeth. 1999. *The Women's Suffrage Movement: A Reference Guide 1866–1928*. London: University College Press.

Embleton, Alice L. 1900. III. *On the Structure and Affinities of Echiurus unicinctus. Transactions of the Linnean Society of London* 8: 77–97.

[12] Pers. Comm. John Maris, who has extensively investigated his family tree, which includes Embletons.

———. 1901. *Goidelia japonica*—A New Entozoic Copepod from Japan, Associated with an Infusorian (*Trichodina*). *Journal of the Linnean Society of London (Zoology)* 28: 211–229.

Hamer, Emily. 2016. *Britannia's Glory. A History of Twentieth Century Lesbians.* London: Bloomsbury Academic.

Hobhouse, Hermione. 2002. *The Crystal Palace and the Great Exhibition: Art, Science and Productive Industry. A History of the Royal Commission for the Exhibition of 1851.* London: Continuum.

MacPherson, W.M. 1895. *Materials for the History of a Church and Priory of Monymusk.* Aberdeen: Taylor and Henderson.

Marner, Serena. 2002. *In Celebration of the Centenary of the Ashmolean Natural History Society of Oxfordshire 1901–2001.* The Ashmolean Natural History Society of Oxfordshire, p. 8. Available Online.

Moore, J.E.S., and Alice L. Embleton. 1906. On the Synapsis in Amphibia. *Proceedings of the Royal Society, B* 77: 555–562.

Ogilvie-Gordon, Maria M. 1897. Microscopic and Systematic Study of Madreporian Fossils. *Philosophical Transactions of the Royal Society* 187B: 83–345.

Poulton, E.B. 1890. *The Colours of Animals.* London: Kegan Paul, Trench and Tubner.

———. 1919. *The Life of Ronald Poulton.* London: Sidgwick & Johnson.

Veley, Lilian J. 1902a. Birds Capturing Butterflies in Flight. *Nature* 65: 299.

———. 1902b. Birds Attacking Butterflies and Moths. *Nature* 65: 392.

Veley, V.H., and Lilian J. Veley. 1898. *The Microorganisms of Faulty Rum.* London: Henry Frowde.

Wachtler, M., and Cynthia V. Burek. 2007. Maria Matilda Ogilvie Gordon (1864–1939): A Scottish Researcher in the Alps. In *The Role of Women in the History of Geology*, ed. Cynthia V. Burek and Bettie Higgs, 305–317. London: Geological Society Special Publications. No. 281.

Walker, C.E., and Alice L. Embleton. 1906. On the Origin of the Sertoli or Foot-cells of the Testis. *Proceedings of the Royal Society, B* 78: 50–52.

———. 1908. Observations on the Nucleoli in the Cells of *Hydra Fusca. Quarterly Journal of Experimental Physiology* 1: 287–290.

CHAPTER 7

Microbiology Learned Through Practice

The three women about to be met in this chapter are distinguished from those in the previous one because, although each of the new trio had enjoyed some post-school education, none had been educated to degree level. Their limited formal education was supplemented in each case, however, by long hours of self-tuition and by guidance from an experienced mentor. And, since each specialised in some aspect of the very young science of microbiology, it is arguable that they were hardly disadvantaged by any lack of education for, in the years when they were 'learning their trade', there was not a great deal of information which could have been gleaned from either text books or university courses. It was the same freedom from accumulated knowledge and longstanding methods which allowed Lilian Veley to move successfully from entomology to microbiology. Microbiology was learned through practice.

Grace Coleridge Frankland, Water-Borne Diseases

Grace was born in 1858, at a time auspicious for any future microbiologist since only eighteen months into the future Louis Pasteur was to deliver his milestone lectures to the Académie des Sciences demonstrating that the spontaneous generation of life was impossible. When Pasteur kept a liquid nutrient medium ('broth') in sealed tubes, spoilage was prevented in tubes that he heated, just as it was when he applied any one of a range of treatments inimical to life. Bacteria were the living germs of life, and of disease.

P. Ayres, *Women and the Natural Sciences in Edwardian Britain*, Palgrave Studies in the History of Science and Technology, https://doi.org/10.1007/978-3-030-46600-8_7

Grace would from her childhood have been familiar with theories of disease for she was the youngest daughter and ninth child of a medical household, her father, Joseph Toynbee, being a doctor specialising in infections of the ear and throat (he once treated Queen Victoria for deafness). Joseph was a liberal by instinct, his beliefs including the need for an extension of suffrage to the working classes. He was one of the first members of the Commons Preservation Society, dedicated to the protection of public spaces (Toynbee 1909). Among Grace's siblings, her dearest brother Arnold (1852–1883) was a fellow of Balliol College, Oxford. An economic historian of exceptional promise, he devoted much of his time and energies to the social welfare of the poorer classes. His promising career was cut tragically short by his death at the age of just thirty-one. This Arnold should not be confused with Grace's nephew, Arnold Joseph (1889–1975), who became a world renowned historian.

Grace's early life, including any schooling, is still shrouded in mystery but letters to and from her brother Arnold show that sometime during 1872–1873, when she was about fourteen years old, she and her sister Mary spent six months in Germany. The archives of Bedford College record that in the academic year 1874–1875 Grace studied Latin, English Literature, Arithmetic, and, naturally, German.

On 17 June 1882, at the age of twenty-three, Grace married Percy Frankland, son of the eminent chemist, Sir Edward Frankland. The couple had known each other for several years for as well as being a friend of Percy's sisters, Grace was a friend of the family of Ellen Grenside, Sir Edward's second wife. Indeed, both the Toynbees and the Grensides lived in Wimbledon, their homes less than two miles apart. Grace's eldest sister, Gertrude ('Miss Toynbee' by the conventions of the day), was a bridesmaid at Ellen and Edward's wedding in 1875 (Russell 1996, 350). Grace and Percy themselves were engaged shortly after Percy returned from Würtzburg, where he had obtained not just a Ph.D. (1880) but had been schooled in the methods of modern German science. By the time of their marriage, Percy was employed as a lecturer in organic chemistry at the Normal School for Science, South Kensington, the future Royal, then Imperial, College. The couple's only child, Percy Edward, was born about two years later.

The young science of microbiology was developing rapidly, in no small part thanks to the work of the German pioneer Robert Koch. In 1881, Koch attended the International Medical Congress in London where he demonstrated his ground breaking methods for culturing bacteria on solid media. In 1882, Sir Edward and Percy set up a private laboratory at Grove House, in London's Notting Hill, for the analysis of drinking and other waters, which inevitably led Percy towards a study of bacteria living in water. In 1884, Koch was again in London, at the International Health Exhibition, this time demonstrating improved plate culturing techniques. By 1885, Percy had adopted and improved Koch's methods: in 1886 he published a description of what is now called a Petri Dish, which was a year before Julius Petri published his own account (Ayres 2005, 100–101).[1]

From about this time Grace's scientific life becomes easier to follow, in spite of her activities being closely entwined with those of her husband. Percy's obituarist, William Garner, noted:

> *Probably in few cases have husband and wife collaborated so effectively and enthusiastically in both research and professional work. On one occasion it was said, 'many women in the past have helped their husbands, but Percy Frankland is the first man who had the chivalry to admit it'.* (Rayner-Canham and Rayner-Canham 2008, 425)

Grace's first two publications appeared in 1887, both of them in the prestigious journals of the Royal Society, and both placing her name before Percy's in the list of authors; in other words, she had done most of the work (Frankland and Frankland 1887a, b). 'Studies of some new micro-organisms obtained from the air', was followed by, 'On some new and typical micro-organisms obtained from water and soil'; the former, the Franklands noted, were typically coccoid (spherical), while the latter were typically bacteroid (rod shaped). The papers describe the growth in culture of a number of bacteria about which little or nothing was previously known. Both tube and plate cultures were employed, together with a range of nutrient media. The Franklands established that a previously problematic group of bacteria could be distinguished according to their ability to reduce nitrate to nitrite, encouraging them and others to make much more extensive studies of bacterial involvement in the nitrifying process and, ultimately, in the whole nitrogen cycle which is so important in agriculture.

It was in 1888 that a series of minor disagreements between Percy and his father culminated in a major falling out, with the result that Percy left

[1] It has subsequently become clear that Grace learned her bacteriology from Percy, and not vice versa.

Grove House to take the chair of chemistry at the University College, Dundee. Father and son had for long had a difficult relationship, stemming from the time when Edward had put huge pressure on a young and unwilling Percy to become a chemist. Far from cooling relations between the men, Percy's removal to Dundee merely escalated the rancour, their disagreements centring on money owed by the father to the son, and the son's borrowing of his father's lecture notes without permission. At this point Grace became involved (in the view of Sir Edward's biographer, Colin Russell, Grace was never one to mince her words and was, in the eyes of some of the family, a contributing factor to the continuing feud). Grace wrote to Ellen complaining that Sir Edward had prevented Percy's election to the Athenaeum club, and had helped deprive him of a coveted fellowship of the Royal Society. In her reply, Ellen tried to persuade Grace that both charges were unfounded. Sadly, reconciliation occurred only after Ellen's death in 1899. The first move being taken by Sir Edward who wrote to his son saying that he hoped the misunderstandings in Grace's letters to Ellen would be withdrawn (Russell 1996, 494).

University College, Dundee, was still a young institution at the time of the Franklands' arrival, having opened its doors only in 1881. From the very outset it was forward looking, its classes being open to both men and women (the first university in Scotland to allow mixed classes). It attracted a group of young and exceptionally able professors. Grace's scientific interest during this period focused increasingly on water borne bacteria and disease, the two often being linked in the real world with poverty and infant mortality. The scientific fruit of the Dundee period was a collaborative two-volume book, *Micro Organisms in Water: their Significance, Identification and Removal* (1894), illustrated with Grace's exquisite drawings. The little quoted sub-title of the book is revealing, telling the reader that it was 'Specially designed for the use of those connected with the sanitary aspects of water-supply'. It was aimed at improving public health, no doubt with one eye on the situation in Dundee.

At this point it is helpful to take a short diversion to look at not only conditions in the city and the responses they elicited among the young professor of the university, but to place those responses in the wider context of University Settlements and of Toynbee Hall.

Dundee was one among several very poor cities in Britain, but it was unique in one respect. Women and children (the latter working half-time) made up the greater part of the workforce for the simple reason that the owners of the textile mills were able to pay them much less than men. In proportion to their populations, Dundee had more slums than London. Mortality rates among babies and infants were some of the highest in Britain, and those children who survived infancy were commonly neglected by their overworked and exhausted mothers (Small 2013, 92–93). In 1888, a group of university professors and well-to-do townspeople set up the Dundee Social Union (DSU), whose aim was to improve the living conditions and aspirations of the poorest of Dundee by encouraging students, university staff, and townspeople to mix with the poor and destitute, giving them practical support.[2] The DSU's principles were basically shared with those of the 'University Settlement' movement, which opened centres in Glasgow, Edinburgh, and other cities in Britain, all modelled on Toynbee Hall, Whitechapel (Bruce 2012, 85–93).[3] Named in honour of Grace's brother Arnold, the Hall was set up in London's East End in 1884 by his friend and admirer, Rev. Samuel Barnett. Barnett's dictum to the young university men who supported his project was to give, not money but themselves.

There were several connections between the Barnetts and the Toynbees. When Henrietta, the Rev. Samuel's future wife, was sixteen, in 1867, she had spent three terms at the same boarding school in Dover as Gertrude and Rachel Toynbee, Grace's older sisters. The school was run by Margaret and Caroline Haddon: the former being married to, and the latter being the mistress of, James Hinton (Creedon 2002; Throesch 2017, 77). Hinton was the aural surgeon who took over Joseph Toynbee's practice when Joseph died in 1866, which suggests it was Hinton's persuasion that led the Toynbee girls to go to a school in Dover.

Samuel Barnett was 'an Oxford man'. In the summer of 1875, two years after he and Henrietta were married, he delighted, therefore, in showing off his *alma mater* to his young wife. In Oxford they joined a party which included not only Gertrude and Rachel Toynbee but their young brother Arnold, then in his first year at Pembroke College (Barnett 1918, 302–303). Thus, began a friendship between the Barnetts and

[2] The founding professors included Thomas Carnelly (Percy Frankland's predecessor in Chemistry), and D'Arcy Wentworth Thompson (Biology). Before moving to Dundee, Thompson was Secretary of the 'The Cambridge Committee for the Study of Social Questions', one of its members being Sydney Vines. See Barnett 1918, vol. I, 34–35.

[3] The Women's University Settlement, at Blackfriars in London, was founded in 1887 by women from Girton and Newnham Colleges, Cambridge, Somerville College and Lady Margaret Hall, Oxford, and Bedford and Royal Holloway Colleges. Women from the London colleges were invited to live-in rent free in exchange for their work in the community.

Arnold which was founded on a common concern for the poor and disadvantaged. Among its many activities, Toynbee Hall organised a series of clubs and societies providing recreation as well as information. Its Natural History Society explored the fungi of Epping Forest and visited the museum of the Essex Field Club, while its many guest lecturers included Lord Avebury, Mrs Millicent Fawcett, A.G. Tansley, and Vernon Blackman—each of whom had associations with Linnaeus' Ladies (Barnett 1918, 357).

One of the young graduates who shared Barnett's way of thinking and who, on leaving Trinity College, Cambridge, was one of the first resident workers at Toynbee Hall, was Edmund Beale Sargant, Ethel and Alice's oldest brother. Viewed by the Barnetts as a naïve angel, Edmund was just as dear to them as Arnold Toynbee (Barnett 1918, 88). Even though they may not have met face to face, it seems improbable that Ethel Sargant and Grace Toynbee were not known to each other through the chatter of their siblings long before they both contributed to the proceedings of The International Congress of Women, 1899 (Chap. 4). Complementing Ethel's brief paper on 'Women in Botanical Science', Grace's much longer paper, 'Women and Bacteriology', was aimed at her audience's natural interest in family health and outlined the contribution bacteriology could make to improved sanitation and hygiene, and to the elimination of the 'consumption' (caused by tuberculosis-carrying bacteria which were present in more than 50% of all milk samples).

To return now to Grace and Percy Frankland; they arrived in Dundee a few months too late to be among the founders of the DSU, but Grace and her husband were soon actively involved—how could she not devote her energy to a cause which it is clear would have been close to her beloved brother's heart? When the Franklands left Dundee six years later, they left 'a large sum of money to be spent quietly and without advertisement to help the poorest children of Dundee' (Garner 1948). One of the professors who was most involved with the DSU was Patrick Geddes, professor of botany but equally well known as a philosopher-sociologist: interestingly, in 1898, Geddes received a donation to the DSU's funds from Alice Sargant, 'to be used for a Trust for the housing of women employed in original literary, artistic and scientific work'.[4]

The Franklands spent six years in Dundee, before Percy moved in 1894 to take the chair of chemistry at Queen's College, Birmingham.

[4] University of Strathclyde Archives: GB 249 T-GED/9/207.

'Togetherness' most aptly captures the spirit of Grace and Percy's relationship. Percy, like Grace, spoke excellent German and in their Dundee years the pair visited various technical institutes in Germany, learning lessons which they hoped would benefit the textile industries of Dundee (a city especially noted for its jute industry). They once rode a 'sociable' bicycle, a side-by-side tandem, from London to Norwich. After moving to Birmingham, the Franklands travelled together throughout Europe and, of course, they published together, among their joint works being a biography of Louis Pasteur (1898).

Grace was by the closing years of the century less actively involved in the laboratory and more involved with scientific journalism, contributing several articles to *Nature* on topics ranging from 'Bacteria and carbonated waters', to 'Dr Yersin and the plague virus', and writing a dozen articles for *Longman's Magazine*. With the encouragement of Longman's, in 1903, she revised and brought together many of these short articles in a popular book, *Bacteria in Daily Life* (Longmans, Green and Co.), its chapters including, 'Sunshine and Life', and 'Milk Dangers and Remedies'.

Befitting her status as a journalist and writer of popular articles, Grace joined the newly formed Lyceum Club in 1904 (Eivor 1905). Although clubs for professional women were becoming popular in the first years of the twentieth century, this was the first to invade the male 'clubland' of Piccadilly; these women were, in their own way, seeking fellowship. An offshoot of the Writers' Club for Ladies (founded 1892) which was notorious for its indifferent food and cramped shabby surroundings In Norfolk Street, the Lyceum boasted (apart from better food) an art gallery, library, bedrooms, and hairdressers. Its founder, Constance Smedley, intended the Lyceum should be for ladies engaged with literature, journalism, art, science, and medicine, who required 'a substantial and dignified *milieu* where [they] could meet editors and other employers and discuss matters as men did in professional clubs' (Keighren 2017).[5] Among those who had taken 'an active interest in the formation of the club [was] Lady Aberdeen…'.[6] The Club's first President was Lady Frances Balfour, a woman who, while not a scientist herself, had connections with the centre of the scientific community by virtue of her being the sister-in-law of both

[5] Lady Frances Balfour, founder of the Liberal Women's Suffrage Society, was a close associate of Millicent Fawcett.

[6] *The Times*, 18 June 1904, p. 12.

(the deceased) Francis Maitland Balfour and of the very much alive Eleanor Sidgwick (née Balfour) (Huffman 2018, 87).

If further reason was needed to persuade Grace that she should join the Lyceum it was that the founder's sister, Ida Smedley, was well known to her as both a chemist (at the time, teaching at Newnham College) and as a campaigner for the rights of women scientists (Fara 2018, 172–174). Ida and her great friend Martha Whitely set up a dining society which met every four months at the Lyceum.

Grace had grown up in a household where liberal values were espoused and learning was prized. In adulthood she embraced the idea that women should be allowed equal participation in scientific and learned societies. As her family had observed, she and her sisters did not hide their opinions (Barnett 1918, 302). How close a 'friend' she was of Mrs Farquharson (Chap. 4) is not known but, as seen previously, in 1900 she let her name be put forward to the Linnean Society as a woman who wished to attend one of its meetings. In spite of that request being turned down, and her being associated with the controversial Mrs Farquharson, Grace was successfully elected to the Linnean in 1904. Her attempts in the same year to join other scientific societies met with mixed success. She was admitted to the Royal Microscopical Society, but when, in October, she and eighteen other female chemists signed a petition requesting that they should be allowed full membership of the Chemical Society they were rejected.

In retirement, the Franklands returned to live in Scotland, where they were able to see more of their old friend D'Arcy Thompson. Finally, completing their togetherness, Grace and Percy died within three weeks of each other in 1946.

Gulielma Lister, Slime Moulds

Gulielma fits perfectly the definition of 'Gentlewoman Scientist'. She was not merely among the first female fellows of the Linnean, so achieving equality with rank-and-file male fellows, but she became a leader of fellows of both sexes when in 1915–1917, and again 1927–1931, she was invited to serve on the Linnean Society's Council. She was further honoured in 1929–1931 when chosen to be its Vice-President.

Gulielma was born and lived her whole eighty-eight years in Leytonstone, Essex, where her father, Arthur, being an enthusiastic member of the Essex Field Club, had encouraged his daughter's interest in the natural environment (Fig. 7.1). Making more or less weekly visits to its Stratford

Fig. 7.1 Gulielma Lister, 1926, or 'Miss Gully' as she was known around the small Dorset town of Lyme Regis. She was President of the Essex Field Club, 1916–1919. (Source: Permission of the Essex Field Club)

Museum as she grew up, Gulielma was increasingly drawn into the activities of the Club,[7] finally becoming, in 1916, its first Lady-President (Anon. 1951). In 1927, she presented to the Club's herbarium a collection of more than 250 plants made by her uncle, (Lord) Joseph Lister, while he was a medical student at University College, London, being taught botany by John Lindley (Lister 1927). Over 100 plants were from Essex, clearly demonstrating the Lister family's long commitment to studying the flora of their native county.

Evidence that Gulielma was held in the highest regard by her peers beyond the Essex Field Club is provided not just by her election to the Council of the Linnean Society but by her involvement in the British Mycological Society; in 1903 she joined that Society and was twice made its President, in 1912 and 1932. Her popularity stemmed from her willingness to share her knowledge and enthusiasm with others, but she would not have been invited to serve so many societies in an officer's role if she

[7] For example, *Essex Naturalist*, 8, 53, 1894.

had not also been a person of wise and sound judgement, respected by both sexes. Her popularity extended beyond the world of science for in the Dorset town of Lyme Regis, where she spent her summer holidays, she was affectionately known by the townspeople as 'Miss Gulie' [pronounced with a hard 'G'].

Gulielma's knowledge of all branches of natural history was said to be encyclopaedic. Her area of special expertise lay in a group whose members show both fungal and animal characteristics, the Mycetozoa (Myxomycetes, or 'Slime Moulds'); historically studied along with the true fungi, these organisms are classified today within the phylum Amoebozoa. While she managed to enjoy one year of formal education at Bedford College in or around 1876, her main biological education came by way of a mentor, her father. Arthur was by profession a wine merchant. A boyhood interest in botany was revived during a large family holiday in Torquay in 1866, when his brother, Joseph (the surgeon), recounted the lessons in botany which he had received many years previously at the hands of Professor Lindley (Scott 1908).[8] Gulielma was six years old at the time. Then, and thereafter, every holiday became an opportunity for father and daughter to botanise. Arthur became fascinated by mosses, fungi, and lichens, but also made collections of butterflies and moths for the amusement of Gulielma and her five siblings.

On retirement from his profitable wine business, Arthur devoted his life to biology, and the Mycetozoa in particular. Made a fellow of the Linnean in 1873, four years later he demonstrated to the Society mycetozoal streaming, that is movement of plasmodia of *Badhamia*. Arthur's knowledge was helped along the way by Professor Isaac Bayley Balfour (Edinburgh) who donated to Arthur his own collection of mycetozoans which he had made while studying with the legendary Anton De Bary in Strasburg. In 1888, the year of his retirement, Arthur published his first paper, in the *Annals of Botany*. The growing Gulielma was always close to her father's side, learning alongside him. Like him, she was a highly skilled artist. When he published his great work, the *Monograph of the Mycetozoa* (1894), drawings for the seventy-eight plates were shared between them. Gulielma's 'superb draughtsmanship' was gladly utilised by other authors for their volumes, as for instance by F.J. Hanbury in his *Illustrated*

[8] Botany (*materia medica*) formed a large part of the curriculum In any course of medicine at the time when Lister trained. Lindley was the first Professor of Botany in the University of London.

Monograph of British Hieracia, and by A. Dallimore and A.B. Jackson in their *Handbook of Coniferae* (Anon. 1951). Later editions of the *Monograph of the Mycetozoa* were the result of her extensive revisions and additions to her father's original text. Although the mycetozoan floras of Epping Forest, and of Switzerland where she holidayed with her father, were studied at first hand, her at-a-distance studies of those of Malaysia, Australasia, and Japan, made her a world renowned authority on the group, amongst the proofs of which was a cherished gift from another Myxomycetozoan-enthusiast, the Emperor of Japan, grateful for the help she had given him. The Emperor sent to her a handsome pair of porcelain vases in a silk-lined case.

In 1871, Arthur and his two brothers bought a large airy house, High Cliff, at Lyme Regis, in Dorset, which would henceforward be where the greater family gathered each summer. Local historian and geologist, Richard Bull, has made a special study of 'The Listers at Lyme' and has concluded that High Cliff, with its twelve bedrooms, laboratory space, extensive gardens, and five or so servants, was an 'open house' for visiting scientists. (Microscopy was part of the Lister family inheritance; in the 1820s Gulielma's grandfather, Joseph Jackson Lister FRS, had effected huge improvements in the design of achromatic lenses, enabling the fine structure of cells and microorganisms to be better resolved). Arthur was a member of the Dorset Field Club and on one occasion invited seventy members to lunch as part of their annual two-day gathering at Lyme.

Dukinfield and Rina Scott were from July 1889 regular visitors to this Quaker household in Lyme. Arthur Lister and Dukinfield were old friends. Gulielma remembered, '...[Scott] gave a living interest to the subject of fossil botany, of which he had such a masterly grasp'. As others had noted before, and probably thinking of her own interactions with Scott, she added, 'His attitudes to younger botanists and their work was especially sympathetic' (Oliver 1935). Scott was not deterred by the fact that, although the Blue Lias beds of Lyme Regis and its neighbour Charmouth had since the 1830s been renowned for their animal fossils following Mary Anning's discoveries of ichthyosaurs and plesiosaurs, the local Blue Lias yielded few plant fossils.[9] He and Rina loved the beauty of the countryside around Lyme and the whole ambience of High Cliff.

[9] The Bournemouth Freshwater bed, not far away, was much richer in plant fossils. Scott was a member of the Bournemouth Natural Science Society, and its President in 1931.

The charms of Lyme Regis and the Listers have another connection with the Linnean Society and that involves Agnes Robertson (Mrs Arber). When the young Agnes was still at the North London Collegiate School, Gulielma came to speak to the school's scientific society. Finding such an exceptionally enthusiastic pupil, Gulielma sent her a little box containing an exquisitely mounted and labelled collection of mycetozoa (Haskins 1999). The pair kept in touch. After Agnes left school she was to encounter Gulielma's future sister-in-law, Dorothea Marryat, when they both studied at Newnham College. Dorothea later married Gulielma's brother, the Cambridge zoologist and plant collector, Joseph Jackson Lister (not to be confused with his grandfather, of the same name).

Agnes and her own daughter Muriel (born 1913) were to be regular visitors to Lyme from 1922 onwards, four years after Agnes' husband had died. Gulielma became an aunt-like figure to Muriel, and guided the young girl's growing interest in geology. It is said that Dorothea, *aka* 'Aunt Dolly', gave £25 to support that first holiday of Agnes and Muriel in Lyme. Aunt Dolly was a well-connected and respected biologist; in Cambridge she had worked under Professor Harry Marshall Ward on plant disease resistance, before being assimilated into William Bateson's team of female geneticists (see Chap. 8), where she worked on a range of problems. Muriel's love of the Listers and how they and Lyme set her on the road to a distinguished career in geology—she was twice President of the Geological Association (founded 1858)—is reflected in her semi-autobiographical work, *Lyme: Landscape with Figures* published in 1988 (Robinson 2007).

It was in the 1880s that Arthur Lister and his daughter started visiting the British Museum (Natural History, BMNH) to inspect its collection of Mycetozoa. Gulielma was to prove a regular and frequent visitor over the next half century, adding to and helping curate the Museum's collection. In the Botanical Gallery she and her father mounted an exhibition of British species, a display augmented by Gulielma's own sketches and drawings. The Museum was to prove important in Gulielma Lister's life in another way, for it was there that she met Annie Lorrain Smith, who was to be one of her dearest friends.

Annie Lorrain Smith, Pillar of the British Museum (Natural History)

Annie was six years older than Gulielma; the early years of her life are not well documented but it seems that in her mid-thirties, around 1888, and after working as a governess in France and then Germany, she somehow joined DH Scott's classes at the Royal College of Science, assiduously filling many notebooks with her neat tidy handwriting and even neater anatomical diagrams. (These may be seen among the archives of the Museum; her use of watercolours suggesting that some diagrams, if not all, were completed or redrawn *after* classes had finished). As seen already, the greatly admired Scott took his students to the BMNH and it is there in the cryptogamic herbarium that Annie found what was to be her home for the next half century. Since women were not admitted to the staff of the Museum, Annie had to be classified as an 'unofficial worker', paid from a 'special fund'; in reality, money was channelled through a bank account set up with Scott's help.[10] Some income was essential since Annie could not fall back on any family wealth; her father was a poorly paid minister of the Church of Scotland in the village of Halfmorton, in Dumfriesshire, two miles north of the Anglo-Scottish border. The family was large and education was clearly prized, as evidenced by the fact that three of her brothers were professors in later life.[11] Inevitably, their education must have been expensive so the family's limited resources would have had to be directed primarily towards the education of the sons, rather than that of any daughters, as was normal for the times. There was precious little money left for Annie's education, or her support in adulthood.

Annie's first work for the Museum was remounting a collection of Anton de Bary's microscope slides recently purchased by the museum. Drawing on the knowledge of microfungi which she acquired in this way, she next undertook the arrangement of displays of microfungi in the Museum's Botanical Gallery. After a spell assisting William Carruthers, consulting botanist to the Royal Agricultural Society, with his studies of fungi affecting the germination of seeds, she was given the responsibility of identifying fungi on plant collections sent to the Museum from

[10] It was not until 1928 that women could be officially employed at the BMNH.

[11] James Lorrain Smith FRS did pioneering work on respiratory physiology in collaboration with John Scott Haldane, the man who was instrumental in persuading the Physiological Society to admit Florence Buchanan and five other women (Chap. 2).

throughout the British Empire—an enormous task given the ever increasing flow from the colonies. Perhaps seeking expert help with her problems, and certainly with great enthusiasm, Annie was one of the first women to join the fledgling British Mycological Society in 1897–1898 (Chap. 2). She was greatly honoured in 1907 when she became the Society's first female President (a position she filled again in 1917), succeeding Arthur Lister. She was by then already a fellow of the Linnean Society.

Annie's interests underwent a major change of direction in 1906 following the death of the Museum's James Crombie, who left unfinished the second volume of his *Monograph of British Lichens*. Finishing Crombie's work, and producing revised and extended versions over the next twenty years, Annie made herself an expert on lichens: indeed, by the time her own textbook, *Lichens*, appeared in 1921, she was a world authority, her book still being regarded as a classic in 1975 when it was reissued. Annie and Gulielma were often together on field excursions of the Essex Field Club, Gulielma serving as the 'recorder' for any Myxomycetes collected, and Annie as the 'recorder' for lichens.[12]

Another connection, or rather a thread of connections, exists between Annie Smith and the Essex Field Club. One of Annie's closest friends at the British Museum was Ethel Sarel Barton, who joined in 1889. They had both been students in DH Scott's botany classes at the Royal College of Science and now Ethel—like Annie, one of the Department of Botany's many officially unpaid female workers—was becoming a leading authority on marine algae, with a string of publications to her name (Britten 1922). In 1902, Ethel married Antony Gepp, a long serving member of the Museum's staff (1886–1927) and, like her, an expert on marine algae. The couple set up home at Kew. Antony Gepp's brother, the Reverend Edward Gepp, was one of the Essex Field Club's most energetic members, occasionally enlisting Antony's help in the identification of difficult botanical specimens collected on the Club's forays. It was Antony who signed the Certificate of Recommendation for both Annie Smith and Gulielma Lister as they sought fellowship of the Linnean. Which raises the question, why was the admirably qualified Ethel Sarel Barton Gepp not among the first of Linnaeus' Ladies? The answer probably relates to the trials of motherhood; after the Gepp's first child was born and died shortly afterwards in 1903, a second was due in July 1905; in 1904 Ethel was probably more

[12] For example, *Essex Naturalist*, 17, 235 (1912).

concerned with a mixture of grief and hope rather any prospect of joining the Linnean Society. She can have had no ill feelings towards the Society for in 1908 she did become a fellow.

There is a charming eye witness account of Annie Smith and Gulielma Lister together at the Linnean.

> *I see her [Miss Lister] a regular attender at meetings of the Linnean Society, entering the lecture room and turning left into the fourth or fifth row from the door, where she would be joined by Miss Annie Lorraine-Smith [sic], and Miss Noel. She [Miss Lister] wore a neat dark tailored suit and had removed her hat in deference to the sexless character of a Fellow. It was an unusual thing then for a lady to remove her hat, but we all took our cue from Miss Lister and did the same.* (Grace Waterhouse FLS [cited by Haskins 1999])

In Mary Creese's *Ladies in the Laboratory?* (p. 37) she writes that Annie '…was a strong supporter of women's demands for full citizenship and equality of opportunity'. It is not known whether her feelings were ever translated into actions, beyond a general willingness to help young naturalists of both sexes. Gulielma, apart from being Chair of the School Nature Study Union for a number of years, actively promoted the cause of women in botany through her work as a Trustee of the Botanical Research Fund from 1917 until her death.

In echoes of the life of Ethel Sargant, Annie Lorrain Smith never married, instead sharing her home for most of her life with a sister whose death in 1933 severely depressed Annie through the last four years of her own life.

* * *

Although the lives of Grace Frankland, Gulielma Lister, and Annie Smith were lived in very different ways, the women were united by a fascination with the new world of microbiology. Another thing they had in common was that they were connected, not merely with each other and the wider group of women who sought advancement in the professions, but with many of the men who were currently leading the biological sciences. The support given by three such men, in particular helping a relatively large number of women to gain access to a laboratory, is explored in the next chapter.

References

Anonymous. 1951. Obituary Notices. Gulielma Lister. *Essex Naturalist* 28: 214.

Ayres, P.G. 2005. *Harry Marshall Ward and the Fungal Thread of Death*. St Paul, MN: American Phytopathological Society.

Barnett, Henrietta. 1918. *Canon Barnett. His Life, Work, and Friends by His Wife*. Vol. I. London: John Murray.

Britten, J. 1922. Ethel Sarel Gepp (1864–1922). *Journal of Botany, British and Foreign* 60: 193–195.

Bruce, Lynn. 2012. *Scottish Settlement Houses from 1886–1934*. Ph.D. thesis, University of Glasgow, pp. 85–93. Available Online.

Creedon, Alison. 2002. A Benevolent Tyrant? The Principles and Practices of Henrietta Barnett (1851–1936), Social Reformer and Founder of Hampstead Garden Suburb. *Women's History Review* 11: 231–252.

Eivor. 1905. En kvinnlig bakteriolog. *Idun* 26: 325–316.

Fara, Patricia. 2018. *A Lab of One's Own. Science and Suffrage in the First World War*. Oxford: Oxford University Press.

Frankland, Grace C., and P.F. Frankland. 1887a. Studies of Some New Micro-organisms Obtained from the Air. *Philosophical Transactions of the Royal Society of London, B* 178: 257–287.

———. 1887b. On Some new and Typical Micro-organisms Obtained from Water and Soil. *Proceedings of the Royal Society of London* 43: 414–418.

Garner, W.E. 1948. Obituary Notice. Percy Faraday Frankland, 1858–1946. *Journal of the Chemical Society (Resumed)* 3: 1996–2005.

Haskins, E.P. 1999. Miss Gulielma Lister F.L.S. Remembered. *Mycologist* 13: 54–56.

Huffman, Joan B. 2018. *Lady Frances. Frances Balfour, Aristrocrat Suffragist*. Kibworth: Matador Press.

Keighren, Innes M. 2017. 'A Royal Geographical Society for Ladies': The Lyceum Club and Ladies Geographical Frontiers in Edwardian London. *The Professional Geographer* 69: 661–669.

Lister, G. 1927. On a Collection of Plants Made by the Lord Lister Between the Years 1844–1848, Many of Them from the Neighbourhood of Upton. *Essex Naturalist* 21: 104–107.

Oliver, F.W. 1935. *Dukinfield Henry Scott, 1954–1934*, 823–837. XLIX: *Annals of Botany*.

Rayner-Canham, Marelene, and G.W. Rayner-Canham. 2008. *Chemistry Was Their Life. Pioneer British Women Chemists, 1880–1949*. London: Imperial College Press.

Robinson, E. 2007. The Influential Muriel Arber: A Personal Reflection. In *The Role of Women in the History of Geology*, ed. Cynthia V. Burek and Bettie Higgs, 287–294. London: Geological Society.

Russell, C.A. 1996. *Edward Frankland. Chemistry, Controversy and Conspiracy in Victorian England*. Cambridge: University Press.

Scott, D.H. 1908. Arthur Lister F.R.S. *Journal of Botany, British and Foreign* xlvi: 331–334.

Small, E. 2013. *Mary Lily Walker. Forgotten Visionary of Dundee*. Dundee: University Press.

Throesch, Elizabeth L. 2017. *Before Einstein: The Fourth Dimension Fin-de-Siecle Literature and Culture*. London: Athena Press.

Toynbee, Gertrude. 1909. *Joseph Toynbee FRS. Aural Surgeon*. London: Henry J. Glaister.

CHAPTER 8

An Unavoidable Need for Male Support

The harsh truth is that women who hoped to build a career in the natural sciences needed male support. It was unavoidable. In finding employment, gaining access to a laboratory, and generally throughout their early careers, women needed the support and guidance of influential and well connected men. They needed supporters as they first attempted to join prestigious scientific societies, because those societies typically required each application to be signed, or sponsored, by a number of their existing members, and all those members were men—at least until women made the first breakthrough, after which they could sponsor each other. And they needed men if they were to get their research recognised and published.

A lucky few were supported by a father or husband, but most needed help from outside their family. Thankfully, as has been seen, they found it in men such as the Liberal peer Lord Avebury, Baron von Richthofen, Sir Archibald Geikie, John Scott Haldane, Norman Lockyer, Edward Poulton, William Ramsey, and many others. For female botanists, no men were more supportive than D.H. Scott of the Jodrell Laboratory, and his friend F.W. Oliver of University College, London, both liberal-minded men, exceptionally well connected not just in the scientific community but also in the arts. They and another great supporter of women, the pioneer geneticist, William Bateson, are the focus of this chapter.

P. Ayres, *Women and the Natural Sciences in Edwardian Britain*,
Palgrave Studies in the History of Science and Technology,
https://doi.org/10.1007/978-3-030-46600-8_8

Dukinfield Henry Scott, from a Famous Family of Architects

I was very glad indeed to hear of my election to the Linnean Society. It is a great thing to have got the first batch of women Fellows elected ...and I feel all women should be grateful to you for having steered the ship into port. Ethel Sargant to DH Scott, 17 December 1904[1]

Scott's name has cropped up many times already, not least because as Botanical Secretary of the Linnean Society in 1904–1905 he was very much involved in the practicalities of admitting women fellows, about half of whom were botanists of one form or another.[2] Scott has already been seen to be pivotal in the scientific achievements and careers of both Ethel Sargant and Margaret Benson. He was one of a small number of men who supported women's careers in practical ways, another was Francis Wall (Frank) Oliver, Scott's erstwhile student and later collaborator and friend. Even when she was long established at Royal Holloway College, Benson used Oliver as a sounding board for many of her botanical ideas, as letters in the college's archives show, and she chose Oliver as the external examiner of her students.

The Scotts, Dukinfield and his wife, Rina, and Oliver and his wife, Mildred Alice, were friends who shared a love of entertaining and socialising, both families noted for a hospitality which frequently underpinned their professional relationships. A relaxed generous nature may come more easily when one is born into wealth, as was Scott. He was the youngest son of the world famous architect, George Gilbert Scott, who designed the Albert Memorial; he was great uncle to the equally distinguished architect Giles Gilbert Scott, designer of Waterloo Bridge, and Liverpool's Anglican Cathedral.[3] Oliver wrote in his friend's obituary:

We botanists may think of Scott as one of ourselves, but he never ceased to belong to the Scotts, a numerous clan, bristling with talent. He was as faithful to them as to us, and loved to circulate to their marriages and gatherings just as he did on scientific occasions. He was a large-hearted man intensely interested in human affairs. Put him alongside a woman of the world with a spice of frivolity

[1] Scott Letters, Archives of the Natural History Museum.

[2] Mycology has traditionally been treated as part of botany.

[3] George Gilbert Scott was the greatest exponent of the Gothic Revival style advocated by John Ruskin.

and you would discover another Scott, equally scintillating, in this different field. He loved a gossip. (F.W. Oliver 1935)

Scott was one of the most distinguished botanists of his time—or of any time—his original papers, books, and services to several societies earning him numerous awards and accolades from his peers. Oliver continued,

> *After graduating in a non-scientific branch (classics) at Oxford he was put to a course of (railway) engineering study, influenced we may suppose by his father…' After his father's death, and at the age of twenty five, he was free to follow his interest in botany, …[he] found his way to Kew, and was advised by Thiselton-Dyer, then Assistant Director, to proceed to Germany, then the centre of inspiration of the new botany.*

After an intensive three-month German language course, Scott entered the laboratory of Julius von Sachs in Würtzburg, where he obtained a Ph.D. On his return from Germany in 1882, he was put in charge of practical classes in botany at University College, London, appointed by Professor Daniel Oliver, Francis' father.[4] Francis was one of Scott's students. In 1885, Scott moved on to the Royal College of Science, South Kensington, where he was Assistant Professor of Botany to T.H. Huxley, and where among his first batch of students was Miss Henderina Klaassen.

To complete this brief account the young Scott's career, while in his first year at the Royal College he gained the agreement of Thiselton-Dyer, by then Director of RBG, Kew, to take his class of advanced students to the Jodrell Laboratory. He must have made a good impression on the Director because in 1892 Dyer invited him to become the Honorary Keeper of the Jodrell, an unpaid post he gladly accepted and held for fourteen years.[5] Each working day he walked the two miles from his home, The Old Palace Richmond, through Kew Gardens to the Jodrell.

In a book about women's struggle for equality it may seem inappropriate to mention Rina's accomplishments as a hostess, but they were judged

[4] In *The Times* of February 1, 1934, Miss Mary Adamson said, 'All women should honour the memory of Dr DH Scott, for he was the first lecturer on Botany at University College who allowed women to enter his class' (cited by Seward 1934). However, in Oliver's obituary of Scott (Oliver 1935) he points out that classes had always been open to both men and women. Most importantly, women perceived Scott as welcoming.

[5] Scott employed three assistants, G.T. Gwilliam, W.C. Worsdell, and L.A. Boodle, each paid from Scott's own pocket (Desmond 1995, 287).

significant at the time and undeniably underpinned the Scotts' pivotal role in the botanical community. Both Rina and Dukinfield were sociable people, who enjoyed entertaining, either at The Old Palace, where they lived while he was Keeper of the Jodrell, or at their large farmhouse, East Oakley House, near Basingstoke, Hampshire, where they retired in 1906. At the latter, 'Foreign botanists met with a warm welcome, and any diffidence they might have felt was soon dispelled by the joyful hospitality of their host and hostess' (Oliver and Seward 1934).

Dukinfield was an ideal Botanical Secretary for the Linnean Society, for he was well-connected and well liked. He managed to avoid involvement in the falling-outs and personal acrimony which attend even the most high-minded scientific institutions. Thus, when Sydney Vines, one of the founders of the *Annals of Botany* in 1887 resigned only two years later, feeling he could no longer tolerate the personal criticisms levelled at him by Thiselton-Dyer, the 'near collapse' of the young journal was averted when Scott took over Vines' heavy responsibilities (Jackson 2015, 10). Scott served the *Annals* for forty-seven years, a sequence ended only by his death in 1934. Similarly, when the proposal that Agnes Arber should be President of Section K at the 1921 meeting of the British Association for the Advancement of Science in Edinburgh was vigorously opposed by the botany Professors of the universities of Edinburgh (Isaac Bayley Balfour) and Glasgow (F.O. Bower), on the grounds that one women should not follow another (they called it 'a botanical gynocracy'), but refused to accept Arthur Tansley as a compromise candidate, it was the well-liked Scott who stepped in to assume the Presidency. He replaced Arber, a woman whose career he had done so much to support—and continued to support (Boney 1995). He was a peace-maker, acceptable to all camps.

Frank Oliver had first-hand experience of Scott's kindness to his students, citing an instance when as a young man he, Oliver, was perplexed by the question of the best angle at which to cut his fossil blocks before observing them under his microscope. 'These things are horribly puzzling' said Scott, 'let's just work it out from the beginning', which he did with the aid of some drawings. 'Yes, now I think I understand', he said. Scott, the vastly experienced palaeobotanist, had gone through all this play acting simply to avoid a display of technical superiority which might humiliate a less experienced colleague (Oliver 1935). Scott's own mentor in palaeobotany was William Crawford Williamson FRS, of the University of Manchester (previously Owens College). On nearing retirement and seeking a successor, Williamson invited both Scott and FO Bower to Manchester in the winter of 1889–1890 (Bower 1938, 68). After three weeks working with Williamson, Scott was converted to palaeobotany, though not to Manchester. (Somewhat later, Helen Klaassen wrote to her brother-in-law, 'It is a good thing for you that you declined the position offered you there. The smoke and the smells there beat everything' (Howell 2005).) Williamson's widow wrote, 'To Dr Scott's intellectual enthusiasm for the subject, to his tactful veneration for its exponent, and his unfailing kindness, my husband owed much of the enjoyment of his last years. [In the last year of his life] '...he once went to a meeting of the Royal Society, Dr Scott was his guardian; and when the tired, nervous face became overstrained, [Dr Scott] fabricated some excuse for bringing his friend away' (Williamson 1896, 214).

The Scotts' move to East Oakley did not signify any reduction in scientific endeavour. Both Dunkinfield and Rina continued their researches and, in her case, film making. They attended meetings of the Linnean and other societies; on relinquishing the Botanical Secretary's role in 1908 he took over the Presidency for the next four years. A constant stream of foreign and British botanists passed through their home and enjoyed there a brief taste of 'country house science'. Microscopes were set out in a room where visitors could continue their studies and there was a well-stocked library. For relaxation, visitors could walk in the large gardens, which included Rina's beloved rose garden, or walk in the surrounding Hampshire countryside where chalk downs mixed with woodlands and agricultural land.

Francis Wall Oliver, from Kew to University College, London

Frank Oliver was, as already seen, a student of Scott's but as the years passed he became his collaborator, friend, and, ultimately, his obituarist. Through their upbringing, both men had an unusually wide view of the world. Scott's was due to the celebrated architects his family spawned and their connections with the great and good of the day. Oliver's originated probably from his father's artistic interests. Their home at Kew received many of the leading artists of the day, such as John Ruskin who in a letter of 1876 referred to Francis' father, Daniel, as 'my botanical friend, good Mr Oliver of Kew', an advisor on matters both taxonomic and anatomic

(Ruskin 1886, 117). Daniel himself was an enthusiastic and talented artist who each summer escaped Kew for the peace and quiet of somewhere, such as Brittany, where he could sketch and paint without distractions.

The Olivers' home for the fifteen years following their marriage in 1896 was 'an old house of the cottage type with a large garden' in The Vale, Chelsea, 'at the heart of an artistic community' (Oliver 1933, 4). Oliver recalled that, 'Facing us lived William De Morgan the potter and novelist, and his wife [Evelyn], a painter in the Pre-Raphaelite manner. This close contact with the De Morgans proved a great influence in both our lives'. The latter remark is intriguing for it may refer to the De Morgans' belief in pacifism (something consonant with Frank's Quaker roots), or their interest in spiritualism, but it could equally refer to their involvement with the suffrage movement. The De Morgans numbered among their many friends Eustace and Lady Frances Balfour—the latter, a leading suffragist (Huffman 2018, 21). Evelyn De Morgan signed the 'Declaration in Favour of Women's Suffrage' in 1889, while William, whose mother, Sophia, was a founder of Bedford College, showed continuing family support for the suffrage movement by serving as Vice President of the 'Men's League for Women's Suffrage' in 1913. (The De Morgans and the Olivers had one more thing in common for William's father, Augustus, like Frank's, had been a professor at UCL—in the former case, of mathematics).

To add to the Oliver family's connections with the world of the arts, Frank's sister, Ethel, was a close friend of the Sapphic poet, Charlotte Mew, and, on Mew's death, her literary executor. It is hardly surprising given such interests that, in writing Scott's obituary, Oliver looked back fondly to a semi-working trip which he and Mildred made with Dukinfield and Rina in 1905 to Paris, a city that was home to not only some leading palaeobotanists but also to unrivalled collections of art and cultural artefacts (Oliver 1935).

The Olivers, just like their friends the Scotts, loved entertaining, and their circles were cosmopolitan. On one occasion, they, the De Morgans, and a third neighbour, the sculptor, Thomas Stirling-Lee, threw open their gardens for a joint party. Carriages were halted in the King's Road and guests walked into the 'unexpected Fairyland' which was The Vale. 'Old Chelsea Pensioners in their scarlet coats guarded the lane which was festooned with glowing lanterns' (Stirling 1922, 318–319). In each garden there was a different sort of musical entertainment, and guests strolled from one to the other.

Of the three phases of Oliver's career, physiological, palaeontological, and ecological, it was in the second and third that he nurtured the careers of some notable female botanists, among them Margaret Benson, Lilian Clarke, Winifred Brenchley, Ethel Thomas, and Marie Stopes.[6]

Famous, or infamous, for her authorship of *Married Love* (1918) and advocacy of birth control, Stopes' first successes in life were as a palaeobotanist. Marie could be said to be a child of the BAAS (Hall 1977, 16), for it was at a meeting of the BAAS that her parents met. The daughter of Charlotte, a strong-minded, suffrage-supporting woman who had been a classics lecturer at Cheltenham Ladies College (Chap. 3), Marie was educated in her teens at North London Collegiate School. There she was strongly influenced by the chemistry taught under the aegis of Sophie Bryant, successor to Miss Buss as headmistress, and as a consequence hoped to study chemistry at UCL (Green 2013, 106–108).[7] When that hope was dashed by the chemistry department, Marie's ambitions in science were rescued by F.W. Oliver who accepted her to study botany. Thereafter, from the time she entered UCL in 1900, her career was closely directed by Oliver, with some later help from D.H. Scott. Starting in 1903 with a paper, 'On the leaf structure of *Cordaites*' (Stopes 1903), (an extinct Carboniferous seed plant) she had published over forty papers and books by 1918, almost all as the single author (Pearson 2005).

Stopes does not qualify as one of 'Linnaeus' Ladies', however, since she was not elected to the Linnean Society until March 1909 (Pearson 2005).[8] One who does qualify, and who is connected with Frank Oliver, is Marian Busk (née Balfour). The daughter of a rich silk merchant (a distant branch of Francis Maitland Balfour's family), she was privately educated before entering Queen's College, from which she graduated in 1883. Unusually, she married while still a student. Marian was a campaigner for women's suffrage, Treasurer of London University's Women's Suffrage Society, and was proficient in German (Balfour 1942). The last is deduced from her inclusion in the short list of names of those who helped Oliver translate into English the two massive volumes of Kerner von Marilaun's classic

[6] For more about Clarke and Brenchley, see Chap. 9.

[7] Dublin-born Sophie Bryant moved to London as a teenager and attended Bedford College. She was the first woman in Britain to be awarded a D.Sc.

[8] Pearson states incorrectly that Stopes became a Fellow in 1904. Her election did not occur until 1st April 1909, see *Proceedings of the Linnean Society*, 1909, 121, 13.

text, *The Natural History of Plants, Their Forms, Growth, Reproduction, and Distribution* (1894–1895).

Marian and her husband Edward were in 1882 among the cofounders of College Hall, Byng Place, an independent hall of residence for the rapidly increasing number of female students studying in the University of London.[9] Among the other founders were Lady Mary Lockyer (née Browne) and her sister, Annie Leigh, both of whom had been educated at Queen's College and both of whom were keen suffragists.[10,11] Marian's connections did not end there, for her brother, Henry Balfour FRS, an archaeologist and the first curator of the Pitt-Rivers Museum in Oxford, was President of the Ashmolean Natural History Society in 1901, the same year in which Lilian Veley was its Secretary (Marner 2001).[12]

In the third phase of Oliver's career he was, as already seen in Chap. 3, a pioneer of field courses for botany students of *both* sexes. Born into a Quaker family, Francis was at heart a non-conformist; in his science he was what today would be called 'a blue-skies thinker'. Among the subjects on which he published were the effects of air pollution on plants, the invasion of estuaries by *Spartina* grass, and the effects of desert warfare on vegetation, all well before those subjects had attracted wider attention. As retirement beckoned, he surprised everyone who knew him by accepting a chair of botany in Cairo University.

An episode in the life of Ethel Thomas illustrates how Frank Oliver, just like Dukinfield Scott, was held in the opinion of his peers to be open-minded and unprejudiced—a peace maker. Educated at University College and at one time research assistant to Ethel Sargant, Thomas joined the staff of Bedford College in 1906 as an assistant lecturer and when the college established a separate Botany Department in 1908 she was made its first Head.[13] It was a time of rapid personal advancement. When Bedford College moved to Regent's Park in 1913 she designed its botanical garden

[9] Edward was knighted in 1901 and was Vice-Chancellor of the University of London from 1905–1907.

[10] Lady Lockyer was one of the earliest female fellows of the Royal Astronomical Society.

[11] In 1888, Annie Leigh Browne, Lady Aberdeen, and Millicent Fawcett were three of the twelve women who established the Society for Promoting the Return of Women as County Councillors.

[12] Marner, Serena. 2001. *In Celebration of the Centenary of the Ashmolean Natural History Society of Oxfordshire, 1901–2001*. pdf online.

[13] Botany had previously been joined with Geology; head of the joint department was Catherine Raisin.

which was laid out adjacent to the South Villa and botany department. Thanks to a bequest of £1000 made in 1909 by Alice Sargant (sister of Ethel, an ex-student of Bedford, one of the College's Reid Trustees, 1892–1896, and member of its Council, 1901–1909), roughly half that sum could be spent on a botanical library and half put aside for a new physiological laboratory. Unfortunately, due to the intervention of war in 1914, hopes for the latter were not realised until 1925, when a new laboratory and glasshouse were opened by Lord Justice Charles Henry Sargant (Ethel and Alice's brother) in a ceremony 'attended by members of the Sargant family' (Tuke 1939, 254).

Thomas' success and good fortune continued. In 1914 the University of London offered a professorship to her, and in 1915 she received both a D.Sc. and a fellowship from her *alma mater*, UCL. By 1916 she was the Secretary of Section K of the BAAS, the first woman to hold that office. Then suddenly everything changed: in March 1916, Bedford College's Council unexpectedly declined to renew her contract, which expired that autumn. In effect, she was peremptorily dismissed; dismissed for reasons that were never made clear.

Old friends were outraged. Ethel Sargant, who was by now living quietly in retirement in Girton village, was re-energised. 'Mrs Scott, Miss Lister and I have drawn up a circular letter for certain botanists [including Agnes Arber] to sign'. The signatories, '...protested at the iniquity of [Bedford] not allowing E.M.T. to know of what she is accused, and of course giving her in consequence no chance of reply'.[14] Frank Oliver took on the role of intermediary and peacemaker. On 16 September, Sargant was able to tell Agnes, '...you will be glad to hear that E.M.T. has compromised with Bedford College on very favourable terms', forwarding to Agnes a copy of the key document.

The real reasons for Thomas' dismissal remain obscure to this day. Ostensibly, they were connected to the Botanical Research Fund (BRF), although there was never any suggestion of financial malpractice, rather a disapproval of a certain dictatorial behaviour and flouting of agreements by Thomas. Briefly, the BRF was Thomas' idea, and it had been strongly supported by Ethel Sargant. The two women recognised that there was a shortage of places to work in London for experienced female botanical

[14] Letter from Sargant to Arber, 13 July 1916. Archives of Girton College, Cambridge. The abbreviations EMT and ENT are both used by friends when referring to Ethel Nancy Miles Thomas.

researchers (a problem at the forefront of Sargant's mind as she decided to leave Reigate), so in 1913 they started to raise funds for an Institute which could accommodate such women. Realising that it would take many years to finance an Institute, small temporary facilities were erected at Bedford College, being completed and equipped by November 1914. However, in 1916, the Council abruptly cancelled its agreement with the fledgling Institute saying various terms had been breached.

Mary Creese in her, *Ladies in the Laboratory*, suggests the cause was, '...serious difficulties in outlook between the outspoken and occasionally inflexible Thomas and Bedford College's more cautious Principal, Margaret Tuke' (Creese 1998, 40).[15] All Thomas' friends concurred, her character was part of the problem: a strong feminist, she was a mixture of ambition, courage, and warm heartedness, yet hasty at times and 'capable of steely coldness' (Delf 1944–1945). Parallels can be drawn with the character of Marian Farquharson and her treatment by the Linnean Society.

In a very private letter to D.H. Scott, Oliver placed much of the blame on Major Leonard Darwin, Chairman of Bedford College.

> *It was a revelation to me that such an individual as L. Darwin existed at all with his meticulous regard for the letter of the thing, whatever it was. The absolute limit! With Miss Thomas at the other end there was trouble every time.*
>
> *...the worst that could be urged against Miss T was carelessness, a lack of consideration and method in her dealings with the Council, and perhaps contempt hardly disguised.*[16]

Ethel Thomas left Bedford College in 1916, at about this time suffering a nervous breakdown. She did however donate £1000 to the BRF in 1917—probably a large part of the compensation she had received from Bedford College. Although the BRF continues until this day, its records for the

[15] Dame Margaret Tuke, BA, MA, was a 'Steamboat Lady', like Lilian Veley.

[16] Letter from Oliver to Scott dated 22 July 1916; from 'Scott Letters' in the archives of the Natural History Museum. Major Darwin was the fourth son of Charles. He served in the Royal Engineers, and was later Liberal Member of Parliament for Lichfield. He was President of the Royal Geographical Society from 1908 to 1911 and President of the Eugenics Society from 1911 to 1928; also, Treasurer of Bedford College in 1910, as it moved to Regents Park (Chap. 5), and Chairman from 1913 to 1920.

period 1916–1918 are missing, most probably having been lost or destroyed by Thomas.

Such a strong character soon bounced back. After various war duties, Ethel Thomas served as acting head of the Botany Department of University College of South Wales, at Cardiff (1918–1919), and then spent many years, ultimately as Head of Botany, at University College, Leicester, where she and her research students obtained several grants from the BRF. That she had any career at all following her dismissal from Bedford owes much to the brokering skills of Frank Oliver and the respect in which he was held. It may have helped Oliver that in his dealings with Margaret Tuke he was talking with another Quaker, someone similarly schooled in traditions of mutual respect and forgiveness.

William Bateson, Cambridge, Genetics, and Mendel

There is a great contrast in character between D.H. Scott and F.W. Oliver, who did so much to support the careers of young female botanists, and a third man, William Bateson, who did likewise, for where they were gentle, conciliatory, and understanding, Bateson was fiery, outspoken, and made many enemies. He was an opportunist but, as he sought personal advantage, he carried with him a number of women whose talent he recognised and whose labours he utilised in his laboratory. Consequently, his working relationships with them were close and, for the most part, mutually beneficial.

Growing up in St John's College, Cambridge, of which his father was the Master, William was surrounded by women who cared passionately about their rights. His mother, Anna, was in 1884 a founding member of the Cambridge Women's Suffrage Association (CWSA), which affiliated to the NUWSS, and was its secretary for its first six years. (Other founder members included Millicent Fawcett and Anna's daughter, also called Anna). Anna Bateson senior was, recalled Millicent Fawcett, '…a very keen supporter of the Women's Movement in all its branches: I used to call her my "best woman"' (Fawcett 1925, 79). Anna senior was a member of Newnham's first College Council and, on the national scene, was a vice-president of the Women's Liberal Federation—with which Laura Embleton's sister was connected. Several of her 'very attractive children' were to become Fawcett's lifelong friends.

William's sister, Mary (born 1865), was a contemporary at Newnham of Fawcett's daughter, Phillipa, the latter a brilliant mathematician and physicist. Mary, like her mother, served as secretary of the CWSA. An older sister, Margaret (born 1860), was a journalist who worked mainly for the *Queen* magazine, where she used her professional skills to promote the cause of suffrage, starting a feature called the 'Women's employment department'.[17] In 1888, she organised campaign meetings for the Women's Suffrage Society in a number of towns. In 1913, she became president of the CWSA. The last sister, Anna junior, was the first member of her brother William's research team after, in 1887, being appointed an assistant in botany in the Balfour Laboratory. Anna did not remain long in that capacity for in 1890 she turned her back on Cambridge to run her own market garden near Bashley in Hampshire. She did not, however, turn her back on the women's movement and became secretary of the New Forest Suffrage Society.

William was an outspoken and often controversial advocate of 'Mendelism', his research in the critical period 1900 to 1910 being directed towards providing empirical evidence that would support the genetic principles laid down some years earlier by an Augustinian friar, Gregor Mendel (who died in 1884), but which had been largely forgotten.[18] In the half century that had elapsed since publication of Darwin's *Origin of Species* even the leading supporters of evolution by natural selection, such as Huxley, Lankester, and Poulton, had felt increasingly frustrated by their own inability to provide any understanding of the mechanisms of inheritance. Too many influential thinkers—many of them not biologists—were advocating forms of an alternative philosophy, Lamarkism (Boulter 2017, 26–29 and 56). The rediscovery of Mendel's work at the turn of the century gave Darwin's supporters an opening they had long hoped for, and Bateson was the man who seized that opportunity.

Bateson did not, however, slavishly accept all of Darwin's argument. He argued that evolution occurred in small jumps, or 'saltations', whereas his opponents favoured a more gradual evolution, in the style of Darwin. One notable public spat occurred at a meeting of the Linnean Society on 18 February 1904, where Bateson's arguments were squashed by his former supervisor, Raphael Weldon.

As a junior member of the Department of Animal Morphology, Bateson always struggled to finance his research in Cambridge, believing that he,

[17] After marriage she was known as Margaret Heitland.

[18] Mendel's work was done in the garden of the Monastery of Brno, now in the Czech Republic, but then part of the German-speaking Austrian Empire.

or perhaps it was Mendelism, was not properly appreciated. Eventually, he left in 1910 to set up the John Innes Research Centre at Malden, Surrey, but before that he put together a sizeable team of female geneticists (a new term). They included not only Edith Saunders and Dorothea Pertz, but also his sister, Anna, his sister-in-law, Florence Durham, and two women already encountered here, Nora Darwin and Dorothea Marryat. There was thus, from approximately 1895 to 1910, what has been described as a 'familial organisation' about Bateson's group (Richmond 2006).[19]

Nora Darwin tells how she, like many of her contemporaries, was attracted to join Bateson's 'family'. Nora was reading botany but, in addition to the normal course lectures, she also attended in 1905 Bateson's 'Bible Class', a series of lectures on the new subject of genetics delivered at five or six o'clock each week in a remote lecture room. Nora found Bateson a brilliant and inspiring lecturer. In addition, he was, of course, a man who had given his time, as secretary of the Committee for Promoting the Admission of Women to Titles of Degrees, to the women's fight for equality with men in the university.

Dorothea Matilda Pertz

When Anna (junior) left Cambridge, her place was taken by her friend and fellow Newnham botanist, Dorothea (Dora) Pertz. Dora came from a family of distinguished academics. Her German father was Royal Librarian in Berlin, and Dora spent several years of her childhood in Germany before her father died. Her maternal grandfather, Leonard Horner, was a President of the Geological Society, and an aunt had married into the family of the geologist Charles Lyell. As a girl, Dora had accompanied Lyell on a visit to Charles Darwin at Downe House; in her old age she remembered with great delight how she had been allowed to stroke the nose of the horse upon which sat the great man (Arber 1939). Following three years at Newnham College (1882–1885), she worked in the Cambridge Botany School with Charles' son, Francis, publishing five papers with him on the movements of plants, and a couple of her own. Simultaneously, she was working on a project with Bateson which centred

[19] Richmond notes on p. 591 that Florence Durham taught at Royal Holloway College from 1893–1899.

on inheritance in *Veronica* (Speedwell). Dora became increasingly concerned, however, that her grasp of chemistry, physics, and mathematics was not sufficient to allow her to continue to contribute original research of quality. Added to this, in 1910 her long-time collaborator, Francis Darwin, retired. The result was that, like her friend Anna Bateson, she gave up her personal research, though in Dora's case she maintained some connection with botany by employing her language skills to index the most important botanical papers emerging from German journals. In the post-war years, she also drew illustrations for a number of publications, including a series on floral morphology written by her friend Edith Saunders. Writing Pertz's obituary, Agnes Arber called her an amateur, 'in the best sense of the word'.

Edith Rebecca Saunders

The woman who held Bateson's group together, providing continuity and, more importantly, academic leadership, was Edith Rebecca Saunders (Fig. 8.1). She was born in Brighton where her father kept an hotel but was orphaned at an early age. The 1881 census shows her, aged fifteen, living with an aunt and uncle in Hockley, Birmingham. Her uncle was a general practitioner and able to send Edith to the nearby Handsworth Ladies College, from which she progressed to Newnham College. Edith obtained a first class 'degree' in the final Tripos examination in 1888, as a result of which she was awarded a Bathurst studentship enabling her to undertake research in the Balfour Biological Laboratory for Women. Her outstanding abilities were recognised in 1889 when she was appointed Director of the Balfour. For fifteen years, until the Balfour closed, she taught and organised practical classes for the women of Newnham and Girton Colleges, in the post-Balfour years moving to the post of Director of Natural Science for Newnham. Edith's students, '…saw for themselves every last bit of evidence the fresh material might yield', while other demonstrators 'toiled far behind her in assiduity and skill' (Godwin 1985, 165).

Her collaboration with Bateson had begun in 1895 when, using seed that he had brought back from Italy, she undertook breeding experiments with the hairy-leafed and glabrous forms of a small alpine plant, *Biscutella laevigata* (the buckler-mustard: Brassicaceae). The two forms were maintained in the progeny and she was able to distinguish what would soon,

Fig. 8.1 Edith Saunders. (Source: Permission of the Principal and Fellows of Newnham College, Cambridge)

when Mendel's work was rediscovered in 1900, be called dominant and recessive characters. Her findings were published in 1898 in the *Proceedings of the Royal Society*. In 1902, Bateson and Saunders reviewed for the Evolution Committee of Royal Society, in the first of a series of five Reports, the extent to which available data on inheritance agreed with Mendel's principles. Saunders' cross-breeding experiments with *Lychnis*, *Atropa*, and *Datura* (funded by a small grant that Bateson had obtained from the Royal Society) were pivotal since they were clearly consistent with Mendel's principles and proved critical pieces of supporting evidence. In their landmark report, Bateson and Saunders used several terms that have since been integrated into the basic language of genetics, terms such as allelomorph (now allele, meaning one of the two or more alternative forms of a gene), homozygous and heterozygous (having, respectively, identical or different alleles at the same locus on homologous chromosomes), and P (parental), F_1 and F_2 (filial) generations.

At the 'Third International Conference 1906 on Genetics', held in London, Bateson used his platform as President to make sure that the delegates were fully aware of the advances of Mendelism, and the contribution to that progress made by his own group. At the conference dinner, a grand banquet at which Saunders was presented with a Banksian Medal of the Royal Horticultural Society, 'for the value and extent of her researches in the physiology of inheritance in plants', Bateson unstintingly recognised the contribution of Saunders to that progress, and the 'vast reservoir' of supporting data which she and the other women in his 'family' had collected.

> *Had it not been for the work of my friends and pupils—first of all by my colleague, Miss Saunders, whose name has been so deservedly honoured tonight—there would have been nothing at all to justify me in speaking of the work of inheritance.* (Cock and Forsdyke 2008, 249)

A woman of 'strong character and steady purpose' (Hamshaw Thomas 1947), of 'upright carriage, masculine jacket and starched collar' (Godwin 1985, 165), '...those who were privileged to know her intimately found her a most loveable character and staunch friend'. There is no record of Edith Saunders ever having been involved in any of the women's organisations which attracted so many of her peers, but her whole life was devoted to empowering female natural scientists.

* * *

Women who wished to break into the male-dominated world of the natural sciences in Edwardian Britain unavoidably needed male support. Many men were not willing to provide such support—some strenuously opposing any change to women's status—but, as seen throughout this book, there were many other men who were willing.

The three featured in this chapter illustrate a general truth; the reasons why men supported women in science were many and various. Each of the three came from a family that had already achieved professional distinction and, particularly in Scott's case, was financially secure. Thus, there was no imperative to 'improve'. A strong religious belief is not mentioned by any of Scott's biographers. Maybe, he simply enjoyed being surrounded by women, he liked their company and felt at ease with them. He was happily married to a woman whose intellect matched his own. In Frank Oliver's case, his Quaker background informed his character and behaviour. He was sociable, generally well liked, and was a natural peace-maker. As well as his family's previously described connections with the arts, his sister, Ethel, two years younger than himself, was close friends from school onwards with the Chick sisters, Edith (botanist), Elsie (Icelandic scholar), and Harriette (bacteriologist, nutritionist, and future Dame of the British Empire) (Ayres 2012, 52–56). Frank grew up in a home where he cannot fail to have been impressed by the bright and intelligent young women who were often its guests. Where Frank Oliver had two sisters (we know little about Winifred), William Bateson had four, three of whom were active supporters of the suffrage movement—as was his mother. From birth, he lived among women who took every opportunity to advance their sex's cause. Such a childhood could either make the adult man strongly pro- or anti-women; in William's case it was the former. He was, moreover, an opportunist who saw that by helping aspiring young female geneticists he helped himself, advancing his own career (there is no recorded evidence that they felt exploited).

The lives of Scott, Oliver, and Bateson illustrate just a few of the factors—sexual attraction, religious belief, mutual advantage—that to greater or lesser extent might affect a man's attitude to working constructively with women. Other male supporters met in earlier chapters may have been motivated by the same factors, and by some of their own; none of those influences would have been significant, however, if the men had not possessed also an overriding belief in the equality of women.

References

Arber, A. 1939. Miss Dorothea FM Pertz. *Nature* 143: 590–591.

Ayres, P.G. 2012. *Shaping Ecology. The Life of Arthur Tansley*. London: Wiley-Blackwell.

Balfour, L. 1942. Lady Busk (1861–1941). *Proceedings of the Linnean Society* 1941–1942: 271–272.

Boney, D. 1995. The Botanical 'Establishment' Closes Ranks: Fifteen Days in January 1921. *The Linnean* 11: 26–37.

Boulter, M. 2017. *Bloomsbury Scientists. Science and Art in the Wake of Darwin*. London: UCL Press.

Bower, F.O. 1938. *Sixty years of Botany in Britain (1875–1935). Impressions of an Eye-Witness.* London: Macmillan.

Cock, A., and D.R. Forsdyke. 2008. *Treasure Your Exceptions: The Science and Life of William Bateson.* New York: Springer.

Creese, Mary. 1998. *Ladies in the Laboratory? American and British Women in Science, 1800–1900.* London: Scarecrow Press.

Delf, Ellen. 1944–1945. Obituary Notice. Dr EN Miles Thomas. *Proceedings of the Linnean Society* 156: 235–236.

Desmond, R. 1995. *Kew. The History of the Royal Botanic Gardens.* London: Harvill Press.

Fawcett, Millicent Garrett. 1925. *What I Remember.* New York: GP Putnam's.

Godwin, H. 1985. *Cambridge and Claire.* Cambridge: University Press.

Green, Stephanie. 2013. *The Public Lives of Charlotte and Marie Stopes.* London: Pickering & Chatto.

Hall, Ruth. 1977. *Marie Stopes. A Biography.* London: Virago.

Hamshaw Thomas, H. 1947. Obituary. Edith Rebecca Saunders. *Proceedings of the Linnean Society* 158: 75–77.

Howell, A.C. 2005. James Lomax (1857–1934): Palaeobotanical Catalyst or Hindrance? In *History of Palaeobotany. Selected Essays,* ed. A.J. Bowden, V. Burek Cynthia, and R. Wilding, 241. Geological Society of London, Special Publications.

Huffman, Joan B. 2018. *Lady Frances. Frances Balfour, Aristrocrat Suffragist.* Kibworth Beauchamp: Troubador.

Jackson, M.B. 2015. One Hundred and Twenty-Five Years of the Annals of Botany. Part 1. The First Fifty Years (1887–1936). *Annals of Botany* 115: 1–18.

Oliver, F.W. 1933. *Letters from Egypt. Mildred Alice Oliver: 1 January 1869–8 October 1932.* London: Euston Press for private circulation. (Sackler Library, University of Oxford).

———. 1935. Dukinfield Henry Scott, 1954–1934. *Annals of Botany* XLIX: 823–837.

Marner, Serena. 2001. *In Celebration of the Centenary of the Ashmolean Natural History Society of Oxfordshire, 1901–2001.* pdf online

Oliver, F.W., and A.C. Seward. 1934. Dukinfield Henry Scott 1854–1934. *Obituary Notices of Fellows of the Royal Society 1*: 205–227.

Pearson, H.L. 2005. Marie Stopes and a Century of Palaeobotany. *Linnean* 21: 25–28.

Richmond, Marsha. 2006. The 'Domestication' of Heredity: The Familial Organisation of Geneticists in Cambridge, 1895–1910. *Journal of the History of Biology* 39: 565–605.

Ruskin, J. 1886. *Proserpina, vol. 1, Studies of Wayside Flowers.* Orpington: George Allen

Seward, A.C. 1934. Dukinfield Henry Scott, 1854–1934. *Obituary Notices of Fellows of the Royal Society* 1: 205–227.

Stirling, A.M.W. 1922. *William De Morgan and His Wife*. New York: Henry Holt and Co.

Stopes, Marie. 1903. On the Leaf Structure of '*Cordaites*'. *New Phytologist* 2: 92–98.

Tuke, Margaret J. 1939. *A History of Bedford College for Women 1849–1939*. Oxford: University Press.

Williamson, W.C. 1896. *Reminiscences of a Yorkshire Naturalist*. Edited by his Wife. London: George Redway.

CHAPTER 9

Diverse Paths to Dentistry, Exploration, and Wildlife Photography

Each of the women met so far has been of independent character and possessed of a mental strength that enabled her to follow a path which, from one perspective, was highly unusual for a woman in Edwardian times. Yet viewed from a different perspective they were conventional, for each made her mark by contributing to science in a way that is still today the norm for scientists of both sexes. Focus now turns to women fellows of the Linnean who made their mark in more diverse ways, who defy easy classification, who were unconventional. While, at the time, they were the natural inhabitants of a society, such as the Linnean, whose interests encompassed a broad spectrum of activities, some of the women would not today be recognised primarily as natural scientists. Thus, these first female fellows ranged from a dentist to a wildlife photographer. And they include two explorers, women who needed exceptional physical as well as mental courage.

Viola Annette Latham, Dentist and Doctor

Viola is unlike any other woman in this book, for she qualified as a dentist. Not content with that, which was a notable feat for the times, she then qualified in medicine too. This remarkably determined woman, always known as 'Vida', was born in Manchester in 1866, one of ten children of a physician, John Latham, and his wife Mary Ann Whalley. There is little clarity about her early years. After some schooling at Norwich High School

P. Ayres, *Women and the Natural Sciences in Edwardian Britain*,
Palgrave Studies in the History of Science and Technology,
https://doi.org/10.1007/978-3-030-46600-8_9

for Girls (a Girls' Public Day School founded in 1875), she was a student at Ellerslie Ladies College in her native Manchester.[1] She then entered Newnham College, Cambridge, though there is no record of her ever having sat the final Tripos examinations. A link between Ellerslie and Newnham may have been afforded by the mathematician Agnes Bell Collier. Six years older than Vida, Collier had attended Ellerslie before studying at Newnham from 1880–1883; she was so outstanding a student that immediately upon graduation she was appointed college lecturer in mathematics. Nothing could be more natural than that she should encourage bright ambitious girls from her old school to apply to Newnham, where she could then oversee their welfare and progress.

The next certainty about Vida's life is that she earned a Master's degree from the University of London in 1889, and that in the years immediately preceding the degree she had become interested in dentistry and clinical research while working in a London dental practice. Her role there is unknown but in 1888 she published in American journals short papers on pain and on tooth anatomy, suggesting that she had seen plenty of both at first hand. As an aside, a *Punch* cartoon of 1879 sums up attitudes to dentistry as a possible occupation for women: the female would-be dentist addresses the suffering male patient, 'It's nearly out but my wrist is so tired that I really must rest a bit'.

Vida was not put off by such attitudes and she set her mind to becoming a qualified dentist, though her life might possibly have taken a different course. In 1887 she had become a Fellow of the Royal Microscopical Society. Although such membership fits readily with her interest in the study of human tissues, intriguingly, in the same year she published a paper in the *Microscopical Society Journal*, pp. 843–844, on a very different subject, 'Mounting mosses'. Is it possible that Mrs Farquharson, a Fellow of the Royal Microscopical Society and a moss enthusiast, was encouraging Vida in this direction? What is clear is that Vida's attitudes were admired by Mrs Farquharson for in her lecture, 'Women's Work in Science'—read to the Women's Institute on her behalf by Mrs Alfred Pollard in 1903—she says that Miss Vida Latham MD DDS, Professor of Histology and Pathology at the Women's Medical College, North Western University, Chicago, shares her view that women should be allowed to enjoy the same 'pleasure of scientific work' as men. A direct personal connection between

[1] Loevy, Hannshore T. and Kowitz, Aletha. 2016. Health Science Pioneer. Vida Latham DDS MD. *Dental World* online.

the two women seems likely, and why, unless encouraged by Mrs Farquharson, would Vida Latham want to become a fellow of the Linnean Society when, by 1905, her interests had moved some way from those of most fellows of the Linnean, and she had already been living and working in the USA for several years?

Determined to become a dentist, and not readily finding an opportunity to study in Britain, she had in October 1884 asked Professor W. Bowman Macleod at the University of Edinburgh for advice. He was reassuring, telling her, 'Your sex does not present any barrier to examination, nor does it carry with it any disqualification to practice'.[2] He also recommended that she should consider studying at the University of Michigan, or at Harvard, both of whose diplomas would be accepted for licensure in Great Britain. Latham chose to attend the University of Michigan where her work was so outstanding that she was allowed to teach as a demonstrator in undergraduate classes in bacteriology, and also in comparative anatomy, while still herself a student. Frustratingly, just as she graduated in 1892, the General Medical Council in Britain withdrew its acceptance of American dental degrees.[3] Unable to use her degree in Britain as she had hoped, Latham moved to permanently to Chicago.

Yet still her interests were not satisfied; still she had further ambitions. After undertaking even more studies, she obtained an MD degree from North Western University in 1895 and launched herself into a career in oral surgery. Thanks to her talent, energy, and ambition she rose quickly through her profession and by 1905 was chair of the stomatology section of the American Medical Association. To its list of fellows the Linnean could thus add a woman who was both a distinguished doctor, and an ambassador in North America.

Emilia Frances Noel, Botanist and Geographer

Although she was the granddaughter of 1st Earl of Gainsborough, Emilia was no token aristocratic fellow for she was by 1905 well on the way to being recognised as both an accomplished botanist and geographer for she

[2] *dent.umich.edu/about-school/sindecuse-museum/vida-annette-latham-1866-1958.*

[3] In 1895, Lilian Lindsay, a product of Miss Buss' school, became the first qualified woman dentist in Britain, graduating from the Edinburgh Dental School after being refused admission to the London dental schools.

had begun, with a trip to Egypt in 1892, the travels that would occupy most of her long life and take her to most parts of the world.

Emilia falls into the group of women who enjoyed higher education, in her case at Somerville College, Oxford, although she is unique among that group in 'graduating' not in a natural science but in history, having studied French and German in her first two years. Emilia was already twenty-six years old when she started at Oxford and it was not the happiest period in her life. Her studies were twice interrupted by bouts of illness, requiring courses to be retaken, and her degree classification was the lowest possible pass—a fourth class. One tutor's report describes her as having 'a mercurial disposition', while another says she was 'anxious minded'.[4] She was a complete contrast to that other Somerville graduate, the high-achieving Lilian Veley (née Gould) (Chap. 6).

While an undergraduate, Emilia did nevertheless belong to the Ashmolean Natural History Society of Oxfordshire (November 1895 to January 1898), so she must have known Lilian Veley, the Secretary of that Society, and probably also encountered Lilian's mentor, Edward Poulton, the Hope Professor of Zoology, who, like Lilian, was heavily involved with the Ashmolean (Bellamy 1908, 479).

From Oxford she went as an advanced student to Swanley College of Horticulture in Kent, a mixed sex establishment where she won prizes for a diary of her garden work and for having the best notebook of advanced botany, and where she was greatly loved by the younger women.[5] Thanks to family wealth, Emilia could enjoy, like Ethel Sargant, the luxury of being slow to 'find herself', taking her time to realise that she was happiest outside the classroom. She collected and drew plants wherever she travelled—most of her drawings are in the possession of the Linnean Society—but, as revealed by fifty-seven rigorously compiled notebooks, covering the years 1892–1937, held by the Royal Geographical Society, she was also a distinguished geographer, meticulously recording observations made in Africa, Australasia, and the Americas (Whiting 1949–1950).

In 1903 Emilia published *Some Wild Flowers of Kashmir* (Nottingham: J & H Bell), a book which must have helped recommend her to the Linnean. After becoming a fellow of the Linnean, she regularly attended meetings of the Society, often in the company of Gulielma Lister and

[4] Student records. Archives of Somerville College, Oxford.

[5] Swanley opened in 1887, for men only. The first five women students were admitted in 1891 and from 1903 only women were admitted.

Annie Lorrain Smith (Chap. 7). Emilia's health—possibly mental as much as bodily—was improved by the challenges of travel, and with increasing distance from the intellectually intense atmosphere of Oxford.

Lilian Suzette Gibbs, Explorer and Geographer

Gibbs' history has much in common with Emilia Noel's for she too studied at Swanley College, in her case from 1899 to 1901. Thereafter she attended the Royal College of Science. Like Noel, she enjoyed private means and was both a botanist and an explorer, embarking on her first collecting trip while still a student. Her speciality was montane vegetation, and she was blessed with both courage and a constitution able to cope with the physiological stresses experienced at extremely high altitudes. She was the first women and the first botanist to climb over 4000 metres to the summit of Mount Kinabalu (Sabah), which she did without native helpers. Gibbs was made a Fellow of the Royal Geographical Society when it finally admitted women in 1919.

The results of Gibbs' expeditions, for example to Southern Rhodesia and to Borneo, were published in the *Journal of the Linnean Society*. In between expeditions she worked at the British Museum, Natural History. The Museum's website comments that she was 'a characterful woman, …noted for her hostessing skills and feminist ideals'. The Linnean Society's obituary notes similarly, 'She was an ardent supporter of women's causes', adding, 'People who did not know her well may have sometimes thought her intolerant from the emphatic manner in which she expressed her views but those who had the privilege of close intimacy found in her the most loyal and tender-hearted of friends' (Digby 1924–1925). Lilian Gibbs was the epitome of a strong woman.

Lilian Jane Clarke, Innovative Teacher

Connections with Frank Oliver and UCL go on and on; this next one involves a pioneering schoolteacher. Brilliant, eccentric, a martinet, Lilian Clarke was prepared to confront her pupils, telling them bluntly, 'Girls this lesson is getting boring—change the pace'. At James Allen's School for Girls in Dulwich, South London, where she began teaching in 1896, she encouraged her pupils to learn by observation rather than from textbooks, believing that discovering for oneself was at the heart of good educational practice (Sanders 2005).[6] And nowhere were those principles better employed than in the 'Botany Gardens' which on appointment she immediately started to develop in the school grounds.

It was a propitious time for a teacher with Lilian's philosophy. In 1899 the Chelsea Physic Garden asked Professor John Farmer, of the Royal College of Science, to report on the provision of [botanical] educational materials for schools. Farmer concluded:

> *Some of the schools around London maintain small botanical gardens of their own, and it would be of great advantage to them to be occasionally supplied with such rooted specimens as can be from time to time spared.* (Sanders 2007)

The Physic Garden welcomed the report and began supplying schools in the London area, although the supply was erratic. Nevertheless, with the support of William Hales, appointed curator of the Chelsea Physic Garden in the year of Farmer's report, 1899, Lilian was able to develop her teaching garden. Originally organised into systematic beds, that is taxonomically, the garden was later reorganised along ecological lines to represent habitats such as heath, bog, salt marsh, and sand dune (Fig. 9.1). In this redevelopment she was advised by Arthur Tansley, the 'Managing Director' of British ecology; it is not known whether they met at the BAAS, whose meetings Lilian attended whenever possible, or at the Linnean Society, of which Lilian was elected a fellow in April 1905, but Tansley was soon a great admirer of Lilian Clarke (Ayres 2012, 139). (He, Farmer, and Oliver were among those who nominated Lilian for a fellowship of the Linnean.) Plants for the garden were collected from their natural habitats, often by the girls themselves. It was a form of the field work pioneered by Oliver and Tansley.

Born in London, Lilian was well placed to take advantage of the new opportunities for female education opening up in the capital. She progressed through Highbury High School for Girls, the Royal College of Science, and UCL. At the age of nineteen, in 1885, her name appears in the public record as the winner of a Gold Medal of The Society of Apothecaries, an award established seven years earlier to encourage young women to study 'General Botany' which, unlike medical botany, was

[6] See also, Sanders 2014. Gardens for Learning: the work of Lilian Clarke and CT Prime in the development of botanical education in South London. *Society for the History of Natural History*. Online.

Fig. 9.1 Pupils 'learning by doing' at the James Allen's Girls' School. The 'Botany Gardens' were the idea of Lilian Clarke. Originally set out as a series of systematic beds, she later changed them (with the guidance of Arthur Tansley) to represent different ecological types found in Britain: heath, bog, salt marsh, sand dunes, etc. (Permission of James Allen's Girls' School, Dulwich)

deemed 'an improving and refining acquirement' (Fogg 1934). In 1893, she graduated from UCL having studied botany under Frank Oliver. After gaining teaching experience at Battersea Polytechnic College, she was appointed at James Allen's Girls' School.

In the years following her fellowship of the Linnean, Lilian was an active member of the South London Botanical Institute, where Annie Lorrain Smith was a member and where another member, Ethel Miles Thomas—who, like Lilian, was educated at a Girls High School in Highbury, north London—brought parties of students from Birkbeck College Natural History Club (Vickery pers. comm.).

Lilian's expertise was well recognised in her lifetime. She was invited to be a member of the BAAS' committee examining 'Teaching Botany in

Schools', presenting as part of the wider report to the 1903 meeting at Southport her own submission on 'School Gardens'. Sitting on the same committee were such heavyweights as Harry Marshall Ward and Albert C. Seward, the then current and future Professors of Botany in the University of Cambridge, as well as Professor John Farmer. In 1917 Lilian was awarded a DSc by the University of London for a thesis on the subject of botanical education. In his Foreword to her book, *Botany as an Experimental Science in the Laboratory and Garden*, published by Oxford University Press in 1935, after her death, Tansley paid tribute to 'her thoroughly sound fundamental ideas, her extremely clear and honest mind, her keen enthusiasm, and her indomitable energy and perseverance'.

Scores of girls passed through Lilian's classes at James Allen's, but one deserves to be picked out, Winifred Brenchley FLS. After a period learning horticulture at Swanley College, she followed Lilian's example by studying at UCL, where she obtained a first class degree in Botany, again under Oliver. Winifred's real distinction however is that she steadily rose in the botanical world to be appointed the first female head of the Botany Department of the world leading Rothamsted Experimental Station.

Mary Russell, née du Caurroy Tribe, The Flying Duchess

Reaching the fifth of the 'brave and unconventional women', a circle is completed for Mary Russell was the woman who gave formal thanks to the Linnean Society as it welcomed its first female fellows (Chap. 4).

Mary had left Cheltenham Ladies College at the age of sixteen, without any academic qualification or mark of achievement, to join her family in India where her father was a British Army Chaplain (Buxton 2008, 33 and 143).[7] There she met, and subsequently married, Lord Herbrand Arthur Russell, a dashing young officer in the Grenadier Guards. He was not only an *aide-de-camp* to the Viceroy of India but was the future 11th Duke of Bedford, heir to Woburn Abbey and numerous estates in central London. In her marriage, Mary was given an exceptionally free reign by her husband, allowing her to pursue her own interests, which included both shooting and photographing wild life. An astonishingly adventurous

[7] On 8 November 1904 Mary Russell returned to the College to help celebrate its Jubilee and to present a portrait of her old headmistress, Dorothea Beale.

women, Mary is generally remembered for two things, her hospitals and her flying.

It had long been her dream to become a hospital nurse, and even after her marriage to one of the richest men in Britain she attended lectures at the London Hospital, Whitechapel, obtaining a range of nursing certificates. With the wellbeing of the Bedfords' estate workers at heart she took over and improved the old Cottage Hospital in Woburn Village in 1898, only to replace it with the purpose-built 'Maryland' hospital in 1903. At the outbreak of WWI, Maryland was not large enough to cope with the huge numbers of wounded soldiers returning from the Western Front so she converted the Riding School and Tennis Court at the Abbey into what became known as the Abbey Hospital. She was not, however, an aloof, distant Duchess for each day she donned a nurse's uniform and worked long hours in the operating theatre. She was especially interested in the new technique of X-raying, making sure that her hospitals had state-of-the-art equipment. Drawing on her expertise as a photographer, she made herself into an expert, if largely self-taught, radiographer.

Mary's involvement with her hospitals continued long into the post-war years, but in those later times she became better known to the press as 'The High-Flying Duchess'. She took up flying in her sixties, establishing world record flight times from Britain to both India and the Cape of Good Hope. She died as she had lived, adventurously. Flying through a snowstorm in February 1937 while trying to find Cambridge, it seems she became disorientated and headed out over the North Sea. Neither she nor her aircraft was ever recovered.

For such a woman, it is probably the case that being one of the first female fellows of a learned society would not have been found on her personal list of major accomplishments, and having to reply to the toast at the celebratory dinner on 24 May 1905 (Chap. 4) would certainly have been a trial because for the last fifty years of her life she had difficulty hearing, eventually becoming completely deaf. (Mary joked that flying drowned out the tinnitus which was her constant companion.) She was though never a woman to shrink from her duty. More importantly, how did she come to be at the dinner at Princes' restaurant? What was her connection with the natural sciences?

In her authorised biography it is said that, among Mary's many accomplishments, she was an expert ornithologist, which, coupled with her title, and her interest in conservation (Chap. 4) would have been sufficient to qualify her for a fellowship. There is some question, however, as to when that expertise in ornithology was acquired. An early biography by Flora ('Jemima') Green, Mary's long-serving companion, suggests that Mary began watching birds in about 1902 (Gore 1938, 16). In contrast, her officially sanctioned biography, by Meriel Buxton, suggests Mary's serious commitment to ornithology post-dated her election to the Linnean. Buxton observes, 'The early days of her passion for ornithology in 1906 coincided with the start of her extensive correspondence with Dr Sydney Long of Norwich', a pioneer conservationist and for twenty-four years Secretary of the Norfolk and Norwich Natural History Society. Her diaries from 1906 to 1914 are packed with notes about the birds she had observed (Buxton 2008, 91–92). In 1910, Mary became one of the first five Honorary Life Members of the British Ornithological Union and, with the suitably named Dr Eagle Clark, she established the Fair Isle Bird Observatory, visiting and staying on the island nine times between 1909 and 1914.

What Do the Remaining Women Tell Us?

Marian Farquharson fought for the right of 'duly qualified' women to be admitted to the Linnean Society and, applying that criterion, it is questionable whether the remaining women were qualified to join a scientific society. While **Constance Sladen (née Anderson)** was the widow of Percy, who had been Zoological Secretary of the Society from 1885–1895, **Sarah Marianne Silver** was merely the daughter of Stephen William Silver, admittedly, one of the Linnean's most active and long-serving fellows. There is no record of either woman having an active interest in the natural sciences, though Constance did have authority in another field. She had published articles on the archaeology of York Minster, Selby Abbey, and Castle Howard, all located in her native Yorkshire (Nichols 2003).

When barriers were being broken down and an ancient society was moving into uncharted waters it was perhaps inevitable that standards of admission were not applied consistently. None of the Officers of the Linnean had worked harder to get women admitted than had the Reverend Thomas Roscoe Stebbing, Zoological Secretary from 1903 to 1907, a man of 'slight physique and a whimsical humour' (Calman 1927). It was he who obtained the Supplementary Charter which allowed the legal changes to be made which finally opened the Linnean's door to women. His wife, **Mary Anne Stebbing** (née Saunders) was the daughter of a well-known botanist and entomologist, William Wilson Saunders, who was instrumental in founding in 1857 the Holmesdale Natural History Club, of which Ethel Sargant was later the Treasurer. On such family grounds alone, like Miss Silver, Mary might possibly have been admitted to the Linnean.

Mary was, however, a woman with a mission—which was to illustrate the British flora in watercolour. Sadly, much of her work was destroyed by a fire in 1881. Nevertheless, some of it remains for us to admire. A volume of her later drawings, 'Curious Flowers and Fruits', is in the art collection of the RBG, Kew. She was a serious and skilled botanical illustrator for her subjects were not idealised; instead, in a style that was unusual for the time, she faithfully recorded every abnormality and infection. Recording the growth of various seeds and cuttings, many of her drawings are accompanied by notes on the history of the specimen displayed. When two albums containing *ca.* 530 of her drawings and watercolours were sold in 2010, their pre-sale value was estimated by two London auction houses to be in excess of £1000.

Mary not only also helped illustrate her husband's book on the Crustaceae but, with the help of Annie Lorrain Smith, she identified the fungi drawn by her sister-in-law, Anne Stebbing, and it is Mary's pencilled notes which appear on many of Anne's own watercolours (Bynum and Bynum 2017). She was certainly no token wife, which is an accusation which might be levelled against **Catherine Crisp (née Howes)**, whose husband, Frank, was Treasurer 1881–1905 and, as seen in Chap. 4, liked to use his position and personal wealth to impose his will on the Linnean. To be fair to Catherine, she does seem to have shared her husband's interest in microscopes and microscopy for she was one of the first women to join the Royal Microscopical Society.

Catherine also shared her husband's love of horticulture, which found full expression in the gardens of their 120-room Gothic Revival-style home, Friar Park, at Henley-on-Thames. The several gardens within the grounds of Friar Park included a Dutch garden, an Elizabethan garden, a Nosegay garden, a Boccaccio garden, and a 'Marian' garden which was stocked with plants associated with the Virgin Mary (Way 2009, 23–24).[8] Around the gardens were distributed Sir Frank's collection of almost one

[8] The Boccaccio garden was modelled on the medieval gardens depicted in Giovanni Boccaccio's *Decameron* and reflected Frank Crisp's interest in such gardens. His collection of illustrations, *Mediaeval Gardens*, was, with the help of his daughter, Catherine Childs Paterson, published posthumously in 1924 by Bodley Head: London. Two volumes.

hundred garden gnomes, the majority of them sheltering within Friar Park's most celebrated feature, a thirty-foot high model of the Matterhorn mountain, the controversy surrounding which involved **Ellen Ann Willmott**.

Ellen was that rare thing, a *female* celebrity in the Edwardian era (Uglow 2004, 181). To mark Queen Victoria's Diamond Jubilee, in 1897, the Royal Horticultural Society inaugurated the Victoria Medal of Honour which it awarded to those who had made an outstanding contribution to horticulture. Among the first sixty recipients there were only two women, Ellen Willmott and her friend Gertrude Jekyll; among aficionados of gardening and garden history their names are as well known today as in their lifetime.

On her seventh birthday, in 1865, Ellen found a cheque for £1000 on her breakfast plate, a gift from a childless godmother (Brown 2000, 125–129). Her father, Frederick, was a solicitor in the City of London, while her mother was from a family of prosperous lace merchants in Aylesbury. It was in 1875 that Frederick bought a country estate at Great Warley, in Essex, twenty-four miles from the City, and there he set about reviving its gardens. The leading landscape gardener of the day, James Backhouse of York, was called in on the occasion of Ellen's twenty-first birthday to build a rock garden, or more accurately a mountain gorge, accompanied by pools, waterfalls, and rock caves.

A picture emerges of a woman who was pretty, intelligent, fashionably dressed and yet almost totally uneducated. Her personal wealth increased even further in her twenties when the same godmother, Helen Tasker, bequeathed to her the equivalent of £5m in today's money. Following a family tour of Europe to celebrate her thirtieth birthday in 1888, the young heiress was able to purchase Le Chateau de Tresserve and its grounds, near Aix-les-Bains, in south-eastern France. Immediately, she set about buying plants, often at inflated prices, to stock her new garden and also Warley Place which became hers when her father died in 1892.

Although up to 100 gardeners were employed at Warley, Ellen enjoyed being a 'hands-on' gardener, rising early, donning her sabots and apron, and doing her own weeding. She was interested in the origins and breeding of irises, a subject on which she collaborated with the Cambridge physiologist, Professor Michael Foster, who had long made the study of this group his hobby.[9] She made friends with Gertrude Jekyll. They had in common more than their interest in horticulture. Both were keen advocates of women's rights—Jekyll especially so—and both were keen photographers, Ellen again enjoying being a 'hands-on' practitioner with her own dark room. When Jekyll wrote, *Roses for English Gardens*, she acknowledged the help of Miss Willmott for providing 'a considerable number of excellent photographs [there were more than 100 in total in the book], and for valuable help in compiling the list of Rose [*sic*] species as garden plants' (Jekyll and Mawley 1902, iv).

Not content with owning two large gardens, in 1905 Ellen committed herself to a third. Called 'Boccanegra' it was at the western end of Italy's Ligurian Riviera, backed by mountains and, most pertinently, barely two miles across the bay from Sir Thomas Hanbury's celebrated garden, 'La Mortola', perched high above the Mediterranean sea. Sir Thomas had made his fortune in the silk trade in China, although he was a member of a Quaker family better known for its involvement in the pharmaceutical trade. The Hanburys were, through business and marriage, connected with another Quaker family, the Allens, and Thomas' brother, Daniel, was a partner in the firm Allen & Hanbury. The brothers were passionate botanists, especially interested in collecting *materia medica*, which was not surprising given their professional leanings. Daniel was a Fellow of the Linnean and gave a lot of his time to reorganising its herbarium. Encouraged, or persuaded by Ellen, Sir Thomas purchased in 1903 a sixty-acre site of farm and mixed woodland at Wisley, in a rural part of Surrey, which he donated to the Royal Horticultural Society, enabling it to move its headquarters from a cramped site in Chiswick to one with almost unlimited potential.

Ellen was constantly on the move, travelling between her three great gardens, pumping money into each of them. An approaching crisis was hastened first by her financial support of Ernest Henry 'Chinese' Wilson's third plant collecting expedition, and second by her publication of a weighty book, *The Genus Rosa* (of which there will be more, shortly), largely at her own expense. Pivotal to her fortunes was a fire at her Chateau de Tresserve, which proved hugely expensive since she had no insurance. As both her friend and her solicitor, Frank Crisp spelled out the seriousness of her situation but Ellen was reluctant in the extreme to listen.

[9] A man of the broadest interests, Foster was also President of the Yorkshire Naturalists Union in 1898 and, in the same year, Secretary of the Royal Society.

Finally, on 22 March 1907, she took a bank loan of £15,000, at 4% interest, secured against Warley Place and several of her other properties. This is not the place to record each step in her deteriorating finances—extensive details are given by Le Liévre (1980)—except to note that at one stage Crisp was lending her money against the security of Boccanegra, and, until his death in 1919, he was rallying Ellen's friends to help her out with either loans or gifts of money. Properties were leased, but this brought only temporary respite, delaying for a short while the day when they would have to be sold. Finally, only Warley Place remained, still magnificent, and still visited each year by members of the Essex Field Club.

After Ellen's death in 1934, a much diminished Warley Place passed through the hands of a succession of private owners. In 1977 it was leased to the Essex Naturalists' Trust as a nature reserve. Nature may have been allowed, thereby, to re-assert its control, bringing disorder where there was once order but, at least, Warley Place was once again cared for.

The Linnean Society's obituary of Ellen spoke of 'the treasure of her gardens, which were always at the disposal of her friends' (Cotton 1934–1935). One such beneficiary was Dr Norman Moore, a close friend of Francis Darwin and one of his father, Charles', most trusted medical advisors.[10] Ellen Willmott generously gave both plants and advice on garden design when Moore and his wife Milicent (Barbara Bodichon's neice) restored their house and lands, Hancox, in Sussex. However, some self-admitted friends of Miss Willmott told a different tale, 'As gardeners go she was not considered generous, and one looked carefully at gift plants for fear they might be fearful spreaders'—which serves further to underline the complexity of her character (Le Liévre 1980, 81).

Ellen's first venture into publishing, a simple descriptive book, *Warley Place in Spring and Summer*, was a great success. It was composed entirely of her own black and white photographs, which she had taken with her large plate camera, and developed and printed herself. A later publication, *The Genus Rosa*, which attempted to be much more serious, was a complete disaster. She devoted a great deal of time—mainly between 1904 and 1909—energy and, ultimately, money to breeding roses. Unfortunately for her an understanding of Mendelian genetics was in its infancy, and in any case the genetics of the genus *Rosa* are exceptionally complex.

[10] When the teenage Moore left his mother and his native Manchester in order to study in the south, his welfare became the concern of his mother's old friend, Barbara Bodichon. (Moore 2010).

Obviously, she could not be blamed for this, although, lacking any scientific training, she might have been wise to seek expert advice from one or more knowledgeable fellows of the Linnean, such as Edith Saunders or William Bateson (Chap. 8). Whether it was because of her lack of understanding, or for some other reason, she failed to take account of what was known at that time about the taxonomy of the genus *Rosa*. Consequently, her directed breeding was always frustrated, leading her to confess '...the difficulties are insurmountable. I have over and over again felt I was nearing some satisfactory classification and then fresh material has set me astray again'.[11] All of which was undisguised in her book, and probably helps explain why, when *The Genus* Rosa appeared in two volumes, in 1910 and 1914, only 260 of the 1000 printed copies were sold. It was an outcome which hurt not just her pride but her pocket since, at great expense, she had commissioned the distinguished botanical illustrator, Alfred Parsons RA, to prepare over forty colour plates for the book.

The incident involving Frank Crisp was all about his reconstruction of the Matterhorn in his garden at Friar Park. His thirty-foot high scale model of the mountain surmounted a maze of chambers and caves, all connected by an underground stream. The mountain was planted with a selection of the 3000 species and varieties of alpine plants which he had collected, many personally, and was surmounted by two caste-iron chamois. Crisp took offence at Reginald Farrer's Preface to EA Bowles' book, *My Garden in Spring* (1914): Farrer was the self-proclaimed authority on rock gardens, and in his Preface he had made a characteristically no-holds-barred attack on the massing of colour in rock gardens. Farrer's remarks may have been directed generally, as his defenders argued, but Crisp thought otherwise, believing Farrer's criticism was aimed directly at him and his rockery. The passage below would certainly encourage such a belief:

> *...here is nothing but colour, laid on as callously in slabs as if from the paint-box of a child. This is a mosaic, there is a gambol in purple and gold: but it is not a rock garden, though tin chamois peer never so frequent from its cliffs upon the passer-by, bewildered with such a glare of expensive magnificence.* (Bowles 1914, vii–viii)

[11] Letter to the geneticist, CC Hurst. Cited by Le Liévre (1980, 111).

Crisp rapidly produced a riposte in the form a small booklet, *Mr EA Bowles and His Garden, a New Parable of the Pharisee and the Publican*, in which he attacked the unfortunate Bowles for what Farrer had written.

Why then did Ellen Willmott stand at the gates of the Chelsea Flower Show in 1914 distributing Crisp's booklet? The explanation involves more than mere friendship and loyalty, it involves a proxy war between the gardening experts of the day. If Farrer was attacking anyone apart from Crisp, it was almost certainly William Robinson and 'the massing of colour'. Robinson, and the firm of James Backhouse, had had the greatest influence on the design of the English rock garden in the years immediately before Farrer began to try to overthrow their ideas with his own book, *The English Rock Garden* (1913). Backhouse had designed both Ellen Willmott's and, in part, Frank Crisp's rock gardens.[12] Robinson, like Crisp, was a loyal friend who was trying desperately to help Ellen to save herself from her financial collapse. At the gates of Chelsea she was thus defending not only Crisp and Robinson, but possibly hoping to wound Bowles, whom her sister, Rose, subsequently said had never properly acknowledged his debt to Ellen.

Far from being the peaceful, restorative activity so often portrayed today, gardening was in Miss Willmott's life the source of great emotional turbulence.

Emma Louisa Turner[13]

A number of the women whose lives have been described through these chapters had, like Ellen Willmott, some expertise in photography, but there were two who were truly pioneers, pushing photography into new areas. One was Rina Scott, whom we have already met, the other was Emma Louisa Turner. Her works were of the very highest standard and in every year from 1901 to 1911, inclusive, they were shown at the annual exhibition of the Royal Photographic Society. In 1914 she again exhibited and, in addition to her 'Waxwing', 'Redshank wading' and 'Coots', she delivered a Lantern Lecture on 'Birds of Loch and Moor'.

Emma's interests in wildlife photography were relatively new when she was elected to the Linnean for they seem to date only from 1900, the year

[12] Willmott's was a sunken ravine leading to a subterrananean fernery, unlike Crisp's mountainous structure (Bisgrove 1990, 234–235).

[13] Often written, incorrectly, as Louise.

she met Richard Kearton. (In their recent biography of Turner, Parry and Greenwood suggest that the two may not have met until May 1901, when Kearton came to speak to the Tunbridge Wells Natural History and Philosophical Society, at whose meetings Turner was often present, and in 1908 herself a lecturer (Parry and Greenwood 2020, 13 and 67). Whatever the date, it was the beginning of a long and warm friendship). Richard and his brother, Cherry, had published in 1898 a ground-breaking book, *With Nature and a Camera* (Cassell). Much of their work was done in the Outer Hebridean islands. Together, the brothers pioneered the use of hides when photographing wildlife, while Cherry was the first person ever to photograph a bird's nest with eggs. In her turn, Emma's 1911 picture of a nesting bittern in Norfolk was the first evidence of their return to the United Kingdom as a breeding bird, following their extinction in the late 1800s (Parry 2011, 7).

Most of Emma's photographs were taken in and around Hickling Broad in east Norfolk. Although small in stature, and described in *The Lady* magazine as quiet and unobtrusive (Anon. 1927), Emma was tough and wiry. In order to get the best close-up photographs, she was quite prepared to partially submerge herself in cold, muddy water. When help was available, she would get someone to cover her in reeds or any other natural material which was to hand and she would then lie motionless for hours in pursuit of what was often just one image; a single image because each time her camera's plate was exposed she had to go through the difficult process of reloading with another, something that was particularly difficult in the field. For several months of each year she lived on a houseboat moored to a small island in the Broad, and there she had a hut built so that she could store her photographic equipment and develop her photographs (Fig. 9.2). Both in her notes which accompanied her photographs and in her book *Broadland Birds* (1924), which drew together her many years work in Norfolk, the love of her subjects which shone from her pictures was matched by her lyrical prose. In 1924–1925, Emma moved her attention to Scolt Head Nature Reserve in north Norfolk where she was the first 'watcher', or warden, on the island and where, to her annoyance, she was labelled by the national press 'the loneliest woman in England'.

Emma was born at Langton Green, near Tunbridge Wells, in 1867; her father was 'in trade' for he was listed in the 1871 census as a 'grocer and draper'. His business was successful enough to allow a governess and servant to be employed in the family home. Her home, whenever she was not

Fig. 9.2 Emma Turner's accommodation while working on the Norfolk Broads. The houseboat on the right, *The Water-Rail*, was her main living accommodation. An island provided a safe anchorage on Hickling Broad and a place for a small hut, in which she had a darkroom and sleeping accommodation for visitors. Her one constant companion was a large dog, which may be seen to the left of the hut

in the field, remained with her father but in 1913 he died. In the same year her very dear niece Enid went to Cambridge to study mathematics (Parry and Greenwood 2020, 11 and 58), and it is probably at that time Emma when moved to Cambridge. Certainly in 1914 her address was, 'The Old Rectory', in Girton village, Cambridge; in other words, she was living at the same address as Ethel Sargant. Whether the two women knew each other from a time when they both lived in the Tunbridge Wells area remains unknown. What is known is that in 1915 Emma departed the Old Rectory, hiring a cottage with 'two and a half' rooms, a kitchen, a bit of a garden and mullioned windows, all for two shillings (10 p) a week.[14] She

[14] James Parry, pers. comm., quoting a letter from Turner to a Rev. Bird dated 15 December 1915.

lived in her cottage during winters, when even she conceded that field work was impossible, and finally in retirement.

Whether she and Sargant parted on good terms—possibly, after a fixed-term tenancy had expired—is not known. That Sargant represented a friendly influence is suggested in another way for, although Emma was not a graduate, she was an honorary member of the British Federation of University Women, a body with which Sargant was heavily involved and finally its president (Chap. 5) (Haines 2001, 310). Emma was a stalwart of the Cambridge Bird Club and became one of its vice-presidents. So popular was she with the undergraduate members that when, near the end of her life, she lost her sight they would come to her house to read to 'Skipper', as they affectionately called her, from the latest books and papers in the world of ornithology.

These last women, whose qualifications for fellowship of a scientific society were questionable even by the comparatively relaxed standards of that era, provide not just a snapshot, a cross-section, of the women knocking at the door of those societies in Edwardian Britain but they broaden our view of both the connections between the women, and their male supporters, and, of course, the education, wealth, and marital status of such women. Patterns may be recognised more easily, if these women are included, as will be seen in the final chapter.

* * *

The last words in this chapter should be left, however, with the acerbic James Britten, editor of the *Journal of Botany*. They show that even, or especially, at the time when doors were first opened to women, not all were equally welcome.

> *At a meeting of the Linnean Society on December 15th the sixteen ladies who had been proposed for election as fellows were, with one exception, elected. The Society is to be congratulated on the forward step it has taken; many of the new Fellows will be an acquisition to its ranks, and are eminently worthy of the honour conferred upon them. The honour is, however, we think, seriously lessened by the addition of some whose only claim to admission to a learned society is the fact of their relationship to existing Fellows. It is greatly to be regretted that the Linnean Society has not some standard of admission more exacting than that of being "attached to the study" of some branch of natural science; in the cases to which we have referred the "attachment" would seem to be of the slightest.* (James Britten FLS 1905).

References

Anon. 1927. *Lady Magazine*, 17 March.

Ayres, P.G. 2012. *Shaping Ecology. The Life of Arthur Tansley*. Oxford: Wiley-Blackwell.

Bellamy, F.A. 1908. *A Historical Account of the Ashmolean Natural History Society of Oxfordshire 1880–1905*. Oxford: The author.

Bisgrove, R. 1990. *The National Trust Book of the English Garden*. London: Viking.

Bowles, E.A. 1914. *My Garden in Spring*. London: TC and EC Jack.

Britten, J. 1905. Book Notes, News, &c. *Journal of Botany, British and Foreign* 43: 37.

Brown, Jane. 2000. *The Pursuit of Paradise. A Social History of Gardens and Gardening*. London: Harper Collins.

Buxton, Meriel. 2008. *The High-Flying Duchess. Mary Du Caurroy Bedford 1865–1937*. Woodperry Books.

Bynum, H., and W.F. Bynum. 2017. *Botanical Sketchbooks*. London: Thames and Hudson.

Calman, W.C. 1927. Obituary Notices of Fellows Deceased. *Proceedings of the Royal Society, B* 101: i–xxxviii.

Cotton, A.D. 1934–1935. Obituary. Miss Ellen Willmott. *Proceedings of the Linnean Society* 147: 195–197.

Digby, Lettice. 1924–1925. Obituary. Miss Lilian Suzette Gibbs. *Proceedings of the Linnean Society* 137: 72–73.

Fogg, E. 1934. Obituaries. Dr. Lilian Clarke. *Proceedings of the Linnean Society* 1933–1934: 150–151.

Gore, J. 1938. *Mary Duchess of Bedford 1865–1937*, 2 vols. London: John Murray.

Haines, Catherine M.C. 2001. *International Women of Science: A Biographical Dictionary to 1950*. Oxford: ABC-CLIO.

Jekyll, Gertrude, and E. Mawley. 1902. *Roses for English Gardens*, From the Country Life Library. London: George Newnes.

Le Liévre, Audrey. 1980. *Miss Willmott of Warley Place. Her Life and Her Gardens*. London: Faber and Faber.

Moore, Charlotte. 2010. *Hancox. A House and a Family*. London: Viking for Penguin Books.

Nichols, D. 2003. A Biography of Percy Sladen (1849–1900). *The Linnean* (Special issue no. 4): 30pp.

Parry, J. 2011. Broadland's Bittern Pioneer. *Tern* (Norfolk Wildlife Trust), Summer ed.

Parry, J., and J. Greenwood. 2020. *Emma Turner. A Life Looking at Birds*. Norwich: Norfolk and Norwich Naturalists' Society.

Sanders, Dawn. 2005. From Radicle to Radical: Biology Education and the First Women Fellows of the Linnean Society of London. *The Linnean* 21: 20–24.

———. 2007. *The Global and the Local: The History of Science and the Cultural Integration of Europe*. Proceedings of the 2nd ICESHS (Cracow, Poland, 6–9 September 2006), ed. M. Kokowski, 606–607. Krakow: Polish Academy of Arts and Sciences.

Uglow, Jenny. 2004. *A Little History of British Gardening*. London: Chatto and Windus.

Way, Twigs. 2009. *Garden Gnomes: A History*. London: Bloomsbury Publishing.

Whiting, M. Muriel. 1949–1950. Obituaries. Miss Emilia Frances Noel. *Proceedings of the Linnean Society* 162: 232.

CHAPTER 10

They Sought Fellowship but Did They Make Good Fellows?

Woman is an amazing creature. She has always largely made the world. In the future she must make it more and more.

—From a letter, written on 21 March 1889 by Lady Margaret Huggins (spectroscopist and astronomer), to Anna Swanwick (a member of Council of Queen's College and one-time Principal of Bedford College, London) ('Correspondence of Lady Huggins to Anna Swanwick', online at catalogue.nli.ie [National Archives of the Library of Ireland].)

In Daphne Glick's history of the National Council of Women of Great Britain, which describes how in 1895 the Council coalesced from many bodies concerned with social reform, the author remarks tellingly that those involved with such bodies were women who were no longer 'prepared to live a life of idleness' (Glick 1995, 3).[1]

That remark describes the female scientists who fought—sometimes successfully, sometimes unsuccessfully—to join scientific societies which up until then had been closed to them. The relative intensity of each woman's fervour is not recorded, but in most it burned bright. These women

[1] The Council's President from 1916 to 1920 was Dame Maria Ogilvie Gordon.

P. Ayres, *Women and the Natural Sciences in Edwardian Britain*,
Palgrave Studies in the History of Science and Technology,
https://doi.org/10.1007/978-3-030-46600-8_10

were mentally and physically strong enough to be in the vanguard of change, while less fortunate women, of equal ability and ambition—whose names we will never know—fell by the wayside because of ill health, or compelling demands on their energies made by families or friends.

The women were determined to lead active, rather than passive, lives. They wanted to have more control over the direction of their own lives than their mothers or grandmothers had over theirs. Beyond their ambition to integrate into the scientific community, they were often also involved in a wider fight for women's rights. The issue of suffrage was closing in around them. It could not be ignored. If not actively involved themselves, many had family or friends who were.

What Had These Women in Common?

The twenty-five women of the Linnean Society, who have been focused upon, have been divided into two groups, 'the scientists' and 'the others' (Appendix 1A and 1B) distinguished by the former having contributed to their chosen science, by way of either research or writing, or both. What is noticeable about 'the scientists' is what made those contributions possible, that is, their education. Most belonged to a generation able to take advantage of improved educational opportunities, such as the Girls' Public Day Schools, Bedford and Royal Holloway Colleges, and, above all, Newnham College, Cambridge. Although some of the women in the latter group (1B) also shared in the new educational opportunities, many were tutored at home, and ended formal education at a relatively young age. Whether by cause or effect, they were less orthodox, developing more diverse interests, such as exploration, photography, and garden design.

Considering both groups, three factors emerge at a high frequency among the twenty-five women: a security that derived from wealth, being unmarried, and living close to London. Interestingly, the last was dismissed as a factor common to the twenty-one women who were the first to join the Geological Society, although when the list of first female fellows was read out by the President of the Geological Society, on 21 May 1919, eight of the nine fellows resided *at the time* in either greater London or Cambridge (the exception was Maria Ogilvie Gordon) (Burek 2009).[2] Clearly linking the female Fellows of the Linnean, there were friendships for, either at first- or at second-hand, many of them knew each other

[2] Of the twenty-one early fellows listed in Burek's Table 6, only five resided outside the zone stretching across London from Cambridge in the north to Reigate in the south.

before their bid for fellowship. Here, there is a point of similarity with an earlier analysis; that of women petitioners to the Chemical Society who, it was found, had often been educated at, or worked within, the same institutions, thereby becoming aware of each other (Chap. 5). Not surprisingly, several of the geologists knew each other well, and had sometimes collaborated in their work, before becoming Fellows of the Geological Society.

Wealth and Security

Linnaeus' Ladies were in most cases born into families that could financially support their education and give them early choices in life. In some cases such financial support extended well into their adulthood. Sometimes wealth was inherited, as in the cases of Willmott and Fraser, but more often fathers were solid, salaried, professional men—lawyers, doctors, or members of the clergy with a good 'living'—all able to support their sons and daughters in comfort.[3] (An exception was Embleton, and possibly also Smith.) Families tended to be large and it is noticeable that, as in the cases of Sargant, Smith, and Frankland, siblings of both sexes often had distinguished careers, albeit not in the natural sciences. A picture emerges of large, prosperous families in which the parents had a liberal, progressive attitude towards the education of their daughters. They were families in which competition between girls and boys was both natural and a spur to greater achievement.

Marital Status

The majority of the twenty-five were single, whether it was by choice—possibly the fact that marriage led to a loss of legal rights and identity (Chap. 1) was an influence—or the lack of a compatible partner is not known. Whatever the reason, those single women were free of the heavy responsibilities expected of wives and mothers in Edwardian times, which would otherwise have restricted their time and energies.[4] Husbands who supported their wives' interests outside the home, such as John Gordon or Percy Frankland, were unusual. Or, more specifically, husbands who gave

[3] In the nineteenth century stipends varied enormously from parish to parish. A few parsons were paid more than their bishops but many more had to supplement their basic stipend by income from glebe lands or teaching.

[4] Even if wives of officers are included, only eleven of the twenty-five were married (see Appendix).

credit to their wives, as did Percy Frankland (or Pierre Curie), were rare. Any form of collaboration with a man carried risks for the woman for, as Hertha Ayrton famously remarked in a letter to the *Westminster Gazette* defending Marie Curie, 'Errors are notoriously hard to kill but an error that ascribes to a man what was actually the work of a woman has more lives than a cat'.[5]

Geography

Although a few of the women were from Scottish families, such as Gordon, Fraser, and Smith, most lived in or close to London, their homes typically clustering within a circle bounded by Cambridge in the north and Reigate and Tunbridge in the south, a distribution similar to that displayed by the first female Fellows of the Geological Society (see footnote 1). There is a striking absence of daughters from families living in the midlands and north of England, Wales, or Scotland, in spite of the ever growing number of lively field clubs, Lit. & Phil. societies, astronomy societies, and so on, in the regions, many of which (but not all) welcomed women as full members. What explains this discrepancy between local and national membership?

London was then, as it is now, the home of almost all national societies—association with the capital has always leant prestige and implied gravitas. There is, however, more to explaining the discrepancy. London led the nation where the provision of good secondary education for girls was concerned. Possibly more importantly, London was then, as it is now, at the heart of the business life of the nation, providing by far the greatest number of jobs for professional men. By the 1870s, railway lines around London were proliferating, thereby facilitating easy commuting to prized addresses in leafy Wimbledon or rural Reigate. While fathers worked in the city, young families could grow up far removed from the polluted atmosphere of central London. Was it then the case that only women from such homes, women who might most easily travel to meetings of the Linnean Society in central London, had sufficient interest to want to join the Society, or was there another reason why women living further from London did not apply? Could it be that there were London-centred networks, of men, who encouraged and smoothed the path of women to fellowships?

[5] *Westminster Gazette*, 14 March 1909.

The People They Knew

In order to become a fellow of the Linnean a women had to have her Certificate of Recommendation signed by at least three men. Patronage and personal connections were therefore vital. Examples of male support have been described in Chap. 8 but here is another example which serves to underline the London-centred web of personal connections—at least among the botanists.

John Farmer, Professor of Botany at the Royal College of Science, South Kensington, signed the Certificate of Recommendation for Margaret Benson, Ethel Sargant, and Ellen Willmott. He could have done the same for several more women. In 1903 the editor of the *New Phytologist*, Arthur Tansley, probably wishing to fill the pages of his fledgling journal, decided to report meetings of the London Botanical Society, 'founded two or three years ago', by Farmer, for 'informed communication' and the 'discussion of preliminary results'.[6] Meetings were held monthly in the Biological Sciences Lecture room of the Royal College of Science. In his reports for 1903, Tansley mentions the participation of a number of women, Margaret Benson, Rosalie Blanche Lulham, Ethel Sargant, Annie Lorrain Smith, and Grace Wigglesworth.[7] Rina Scott demonstrated her time-lapse photographs and sessions were chaired by, amongst others, DH Scott and FW Oliver. Tansley's reports thus constitute a roll call of names already familiar from these pages, to which could be added, in 1904, Marie Stopes, Ethel Thomas, and Agnes Roberston. Farmer and to an extent Tansley were facilitating professional and social interactions between the women, and between the women and their male peers. But the meeting place was in London.

Ethel Sargant may have had enough wealth to furnish her own private laboratory but her work there would probably have gone unrecognised if she had not first worked at the Jodrell Laboratory. Like Benson, she had the closest connections with London-based DH Scott, and also FW Oliver, being supported and advised by these men from the time when she and Benson took their first tentative steps into the botanical world, through to

[6] Not to be confused with earlier societies having a similar name.

[7] Grace Wigglesworth graduated from the (Victoria) University of Manchester in 1903, and then took an M.Sc. in palaeobotany. Which raises the question: Did she travel to London specifically for a meeting (unlikely), or was she in London learning from the palaeobotanists Oliver and DH Scott (more likely)? Grace later enjoyed a long a distinguished career as Assistant Keeper of Botany at Manchester Museum.

a time, many years later, when the women were themselves recognised as part of the botanical establishment. The young Helen Fraser was a favourite of George Murray, and at critical stages in her early career she was also helped by Vernon Blackman; again, both men were London-based. Saunders and Pertz would probably have had no career at all if it had not been for the head start given to them by William Bateson in Cambridge. In spite of rivalries between botanical groups, most notably between staff at Kew and those at the British Museum (NH), experienced botanists and also students moved between those two establishments, and similarly between them and University College, or the Royal College of Science (Imperial College). To say that 'everyone knew everyone' would scarcely be an exaggeration, and no one had a better overall view of the community, or was better respected, than the Botanical Secretary of the Linnean Society, DH Scott. These examples of support all involve botanists, which is not surprising when botanists constituted by far the largest group among Linnaeus' Ladies; however, support was not confined to botanists as demonstrated by the encouragement and practical help that the entomologist EB Poulton and others with Oxford connections gave to Lilian Veley.

Vital connections could be made via family members. Thus, one of Grace Frankland's supporters was Lord Avebury, a member of the exclusive X-Club which included also Edward Frankland, Grace's father-in-law.[8] (When Grace's research papers were communicated to the Royal Society, the fellows who did so were two other X-Club members, E Ray Lankester and the club's founder, TH Huxley). Eleanor Sidgwick's work in establishing Newnham College's reputation as a home for young female scientists was facilitated by her being born a member of the exceptionally well-connected Balfour family.

A shortage of good schools for girls was a major factor holding back the progress of women in the nineteenth century. The Endowed Schools Act of 1869 led not only to the foundation in 1872 of the non-denominational Girls' Public Day Schools (GPDS) but to those, such as the Anglican Church Schools Company (founded 1882),[9] inspired by the GPDS; by 1900 there were over 160 grammar and high schools for girls (Watts 2007, 99).

[8] Another member of the X-Club was Edward Busk, husband of Marian.

[9] Miss Beale and Miss Buss were soon appointed to its Council. www.churchhigh.me.uk/school-history/background-of-the-church-schools-company-limited.

Schools set up by the Quakers (The Society of Friends), such as the co-educational Ackworth School in Pontefract (1779), and the girls-only Mount School in York (1785), were particularly influential.[10] Quaker involvement in education and learning extended beyond schools and into the scientific establishment. Thus, when examining the connections and networks which may have affected the position of women vis-à-vis the Linnean Society in 1905, the role of Quakers has to be considered because they specialised in the art of networking, in working mutualistically to attain success. Analysing links between the Quakers and the Royal Society, John Brooke and Geoffrey Cantor concluded that not only were Quakers over-represented in the Royal but, among all Quaker fellows elected up to 1900, botany was by far the best represented discipline (Brooke and Cantor 1998, 302–303 and 312–313). Among the botanists were Daniel Hanbury, the Listers (Joseph, Arthur, and John Joseph), and Daniel Oliver. Daniel's son, Francis Wall Oliver, was educated at Bootham School in York, a Quaker school, as was John Gilbert Baker, Keeper of the Herbarium at RBG, Kew, 1890–1899 (where he succeeded Daniel Oliver), who was a nominator for Gulielma Lister's fellowship of the Linnean. Emily Berridge taught for two years in an unnamed girls school in York, most probably at the Mount School, suggesting that she too may have been a Quaker. Brooke and Cantor also comment that within the Linnean Society there were many Quakers, often seedsmen and amateur botanists, albeit without any connection to the Royal Society.

One easily recognised subgroup of Quaker botanists was the apothecaries. Numbered among them were members of the Hanbury family, and their business partners, the Allens. Daniel Hanbury FRS, FLS, spent much of his time travelling the world seeking *materia medica*. On one expedition, to the Holy Land in 1860, he enjoyed the company of his close friend Joseph Hooker, the Director of the RBG, Kew. In spite of Daniel's continual globetrotting, he still found time to serve on the Council of the Linnean and to organise its herbarium (Locke 1916, vol. II, 296). His brother, Sir Thomas, was a friend of Ellen Willmott and was persuaded by her to give the sixty acres of land to the Royal Horticultural Society which now constitute its home at Wisley Gardens (Chap. 9). Their cousin, Frederick Janson Hanbury FLS, is recorded as being present in the same

[10] Both were set up with involvement from the Tuke family, a descendent of which was Margaret Tuke, Principal of Bedford College (1906–1929), see Chap. 8, Newnham College graduate and fellow.

excursions of the Essex Field Club as EM Holmes, an active member of the Linnean and friend and correspondent of Marian Farquharson. Frederick's book, *British Hieracia* (Hawkweeds), was illustrated by Gulielma Lister. Clearly, Quaker networks reached far to connect Linnaeus' Ladies.

It should not be forgotten that Bedford College had strong Unitarian roots and support, as did the Manchester Literary and Philosophical Society, for the Unitarians, like the Quakers, were actively committed to widening the availability of education, especially to women (Watts 2007, 102 and 120; Brown 2011).[11]

Continuing with the subject of patronage, women increasingly benefitted from the help of other women. Friendships might originate in shared experiences, such as the excursion of a field club, or they might have their origins in shared research interests. Through friendship, the women could validate each other's worth, boosting morale when self-doubt crept in, or giving practical advice that would help another woman's career at a critical point. Ethel Sargant in particular (aged forty-two in 1905), but also Benson (forty-six) and Saunders (forty), encouraged younger women, such as Berridge (thirty-three) and Fraser (twenty-six), Thomas (twenty-eight), Robertson (twenty-five), and Stopes (twenty-four) in their careers. Once admitted to the Linnean, the first female fellows were able to sponsor other women, though in the case of Mrs Farquharson's second application it was none of those just mentioned but rather Catherine Crisp, Grace Frankland, and Ellen Willmott who were among the six fellows who recommended her. The group of younger women had a special respect for Sargant and may have adopted her disapproval of Mrs Farquharson, arguing like Sargant that those whose *scientific* credentials were lightweight should not be admitted to scientific and learned societies.

Looking beyond Linnaeus' Ladies, Lady Warwick was important in opening up new opportunities for women, combining as she did the organising of meetings which gave women a voice with the practical founding of Studley College, which opened new opportunities for women interested in agriculture and horticulture. Perhaps the grandest example of a patron in the present context, though, is Lady Aberdeen. Her Presidency of the International Congress of Women, and her role in organising the

[11] Unitarians rejected the notions of original sin and the fundamental depravity of man, the blame for which were placed upon women.

1899 meeting in London, including the session on 'Women in the Biological Sciences', have already been described (Chap. 3). Her connections, whether by birth or by marriage, and her friendship with men such as the several times prime minister, William Gladstone, were impeccable. Under her illustrious leadership, the International Congress was able to attract an impressive list of speakers ranging from a future prime minister, Ramsey MacDonald, to the Quaker philanthropist, Joseph Rowntree. Elizabeth Garrett Anderson and Emily Davies were further high-profile contributors. Social events included a reception held at Stafford House jointly organised by the Marchioness of Sutherland and Lady Aberdeen. Many of the delegates were presented to Queen Victoria at Windsor (Glick 1995, 163).

Lady Aberdeen helped make men of influence aware of the activities and achievements of women in a great swathe of endeavours, not least the natural sciences. In organising both the programme and social events at the Congress, she could draw upon experiences gained in Canada, not simply as the wife of the Governor General, but as a host when the British Association for the Advancement of Science had visited Toronto in 1897. Regarding the latter, it was said, 'Lady Aberdeen's arrangements for her reception choreographed the scientific "grandees" and dictated who had access to them and where' (Higgitt and Withers 2008).

When women were agitating to join the Chemical Society, in 1909, one male member circulated a letter accusing the women of being linked to the suffrage movement. Thirty-one women, led by Ida Smedley and Martha Whitely, coordinated an indignant riposte; they were united, they wrote, not politically but by a love of chemistry (Fara 2018, 180). Most male members of the Linnean Society were clearly more broad minded and tolerant for, although support for female suffrage was sometimes qualified among Linnaeus' Ladies, most of the women they elected were undisguisedly enthusiastic, often being active supporters of the principles of equality in all realms. Some of the women put their energies into the immediate task of improving the condition of working women less fortunate than themselves, while others were directly involved in campaigning for what proved to be the longer-term goal of suffrage. However, since the suffrage movement was still young in 1904–1905, although gaining strength, the caveat has to be added that the women's greatest involvement was typically in the years *after* they became Lady-Fellows.

Were They Good Fellows?

Having been allowed to join the Linnean in the face of significant male resistance, many of its first female fellows set about playing an active part in its meetings, not only satisfying themselves but vindicating their admission, and repaying their loyal supporters. No registers of attendance at meetings were taken but a smattering of comments show that women such as Sargant, Lister, Noel, and Stebbing regularly attended the Society's events. Remarkably, in little more than twelve months from obtaining her fellowship, and happy to overlook the stuffiness of its meetings (Chap. 5), Sargant was in 1906 elected to the Council of the Society—its first women member—proof that women in general, and Sargant in particular, had quickly won the respect of male members. Saunders served on the Linnean's Council from 1910 to 1915, and was Vice-President from 1912 to 1913. Lister served twice on the Council, from 1915 to 1917 and from 1927 to 1931; during the latter spell she was Vice-President from 1929 to 1931.

The mood within other scientific societies was changing too, helped by the example of Linnaeus' Ladies. Both Sargant (1913) and Saunders (1920) were invited to be President of Section K (Botany) of an annual meeting of the BAAS, though in Saunders' case she caused old attitudes to resurface. She nominated Agnes Arber to succeed her at the 1921 meeting in Edinburgh. It was the first time that the BAAS had visited Scotland in twenty years and Scotland's senior professors of botany, Isaac Bayley Balfour in Edinburgh and FO Bower in Glasgow, were outraged that they had not been consulted over the choice of president, as described in Chap. 8. They found a powerful ally in Cambridge's AC Seward who, like them, objected to one female president being followed by another, the so-called 'botanical gynocracy' (Boney 1995). After considerable wrangling, Mrs Arber stood aside and the well-respected DH Scott assumed the role of president, an outcome that was acceptable to, if not welcomed by, both sexes.

When William Bateson founded the Genetics Society in 1919, with the support of Saunders, 20% of its first members were women—an exceptionally high proportion for a scientific society—reflecting Bateson's strong commitment to involving women in his researches (Lewis 1969, 2).[12] Edith Saunders was soon one of its foremost members, eventually in

[12] Cited by Richmond (2006).

1936–1938 becoming its fourth, and first female, President. And, not to be forgotten, Annie Lorrain Smith was chosen to be President of the British Mycological Society in 1907, and again in 1917 (as was Helen Gwynne-Vaughan in 1928).

In March 1905, very shortly after becoming a fellow, Rina Scott demonstrated her moving pictures at a meeting of the Linnean Society (see Chap. 5), as she did similarly in 1907. Other ladies contributed to meetings but, at first sight maybe surprisingly, they contributed little to the Society's journals. The explanation may be that each edition of the *Journal of the Linnean Society (Botany)* contained a few exceedingly long papers. For example, Lillian Gibbs' 1914 paper on the flora of Mount Kinabalu and the highlands of North Borneo comprised 56 pages of general text followed by more than 180 pages of species descriptions. The formats of the *Annals of Botany* and the *New Phytologist* were much more suitable for the shorter papers, reporting laboratory studies, which were more typical of what most of the research-active ladies were engaged in. In 1905 the *Annals* published five 'Articles' or 'Notes' written by women (approximately 16% of the total), all familiar names: Sargant, Robertson, Berridge, Stopes, Fraser, and Pertz. In 1914, the last year before WWI distorted the figures, the *Annals* published twelve 'Articles' or 'Notes' by female authors (approximately 28% of the total), not all of them Fellows of the Linnean.[13] The *Journal of the Linnean Society (Zoology)* did feature much shorter papers; however, excepting those by Embleton (1901, 1903), Warburton and Embleton (1902), and Veley (1905), there was a notable absence of papers written by women, that paucity probably reflecting the comparative rarity of female zoologists within and without the Linnean.

The subject of journals requires that male patronage has to be mentioned one final time. Editors of journals wielded enormous power, for they could decide what to publish, or rather who to publish. DH Scott at the *Annals of Botany* and AG Tansley at the *New Phytologist* were gentle, constructive editors, who encouraged female authors, helping them fine-tune their papers.[14] Other editors, such as James Britten, editor of the *Journal of Botany, British and Foreign* for more than forty-five years, had a

[13] Although the proportion of papers published by women increased steadily over the next half-century, it was not until 1979 that the *Annals* appointed its first female editor, Gillian Thorne.

[14] Tansley included in the first two volumes of *New Phytologist* (1902-3) papers by Margaret Benson, Edith Chick, Rose Jordan, Gulielma Lister, Ethel Sargant, Rina Scott, Annie Smith, Marie Stopes, and Grace Wigglesworth.

different approach and, in Britten's case, a fearsome reputation among potential authors of both sexes. Britten was infamous for his pungent editorial comments, which would have scared off all but the most confident potential authors; not surprisingly, in the *Journal of Botany* between 1905 and 1914 the proportion of all published papers which were written by women hovered between only 3% and 5%.

The Equals of Men

Women in the Edwardian era were increasingly contributors to the natural sciences. Thanks to improved educational opportunities, both in schools and in colleges, they could work and mix socially with male scientists at a level and frequency previously impossible.

The women soon demonstrated that they could contribute to intellectual discussion and, through talks and published papers, add their own original research to the fund of knowledge which constituted the natural sciences. And they proved too that they could perfectly well take on responsible administrative roles within a learned Society. What happened was evolution, rather than revolution. There was no flood of women into the Linnean. The practices of the Society changed little, if at all, because women had joined. That, however, was the whole point: in what they could put into, and what they could take out of a learned society, women were no better or no worse than men, just equal. Men could safely and profitably share what had previously been *their* Society.

There was a stage in the evolution of feminist theory—mostly during the 1970s—when it was important to recover the accomplishments of great women scientists, from Hypatia of Alexandria to Marie Curie. This was 'in order to counter the notion that women simply cannot do science', and also 'to create role models for young women entering science—"female Einsteins"' (Schiebinger 1999, 22). That phase has long passed. As Claire Jones and Sue Hawkins point out in their editors' introduction to the Royal Society's Special Issue (2014), *Women and Science*, if the only women scientists written about are 'heroines', in some ways special or exceptional, the contribution of the ordinary women of science goes unrecognised. This is not only unfair but it is likely to deter young women of today from pursuing a career in science (Jones and Hawkins 2015).

Those women whose lives have been highlighted in this book were, for the most part, good rather than outstanding scientists. In other words, they were just like most male scientists, very few of whom in either that day or this achieve any sort of individual distinction or lasting fame. Only Ethel Sargant's protegée, Agnes Arber (née Robertson), earned the highest accolade in British science, a fellowship of the Royal Society. The others are not celebrated here for particular discoveries which they made. Instead, they are celebrated for what they collectively represent, which is a vanguard. They were the elder sisters who forged a pathway that their younger sisters might follow.

References

Boney, D. 1995. The Botanical 'Establishment' Closes Ranks: Fifteen Days in January 1921. *The Linnean* 11: 26–37.

Brooke, J., and G. Cantor. 1998. *Reconstructing Nature. The Engagement of Science and Religion*. Edinburgh: T & T Clark.

Brown, Megan K. 2011. "A College for Women, or Something Like It": Bedford College and the Women's Higher Education Movement, 1849–1900. Dissertations and Theses University of Portland. Paper 209. Available online.

Burek, Cynthia V. 2009. The First Female Fellows and the Status of Women in the Geological Society of London. *Geological Society, Special Publications* 317: 373–407.

Embleton, Alice L. 1901. *Goidelia japonica*—a New Entozoic Copepod from Japan, associated with an Infusorian (*Trichodina*). *Journal of the Linnean Society (Zoology)* 28: 211–229.

———. 1903. *Cerataphis lataniae*, a peculiar aphid. *Journal of the Linnean Society (Zoology)* 29: 90–107.

Fara, Patricia. 2018. *A Lab of One's Own. Science and Suffrage in the First World War*. Oxford: Oxford University Press.

Glick, Daphne. 1995. *The National Council of Women of Great Britain. The First One Hundred Years, 1895–1995*. London: The National Council of Women of Great Britain.

Higgitt, Rebekah, and C.W.J. Withers. 2008. Science and Sociability, Women as Audience at the British Association for the Advancement of Science, 1831–1901. *Isis* 99: 1–27.

Jones, Claire G., and Sue Hawkins. 2015. Women and Science. *Notes and Records of the Royal Society* 69: 5–9.

Lewis, D. 1969. The Genetical Society—The First Fifty Years. In *Fifty Years of Genetics. Proceedings of a Symposium held at the 160th Meeting of the Genetical Society on the 50th Anniversary of Its Founding*, ed. John Jinks, 1–7. Edinburgh: Oliver and Boyd.

Locke, A. Audrey. 1916. *The Hanbury Family*. London: Arthur L Humphreys.

Richmond, Marsha L. 2006. The 'Domestication' of Heredity: The Familial Organisation of Geneticists in Cambridge, 1895–1910. *Journal of the History of Biology* 39: 565–605.

Schiebinger, Londa. 1999. *Has Feminism Changed Science?* Cambridge, MA: Harvard University Press.

Veley, L. 1905. A further contribution to the study of Pelomyxa palustris (Greeff). *Journal of the Linnean Society (Zoology)* 29: 374–395.

Warburton, C., and Embleton, Alice L. 1902. The life history of the black-currant gall-mite, Eriophyses (Phytoptis) ribis. Westwood. *Journal of the Linnean Society (Zoology)* 28: 366–378.

Watts, Ruth. 2007. *Women in Science. A Social and Cultural History.* London: Routledge.

Appendix: The First Female Fellows of the Linnean Society

1(A) The scientists

Name (unmarried in 1905 unless shown)	*Age (±1) in 1905*	*Education*
Benson, Margaret	**46** (1859–1936)	Newnham College, Cambridge; University College, London, B.Sc., D.Sc.
Berridge, Emily Mary	**33** (1872–1947)	Dulwich High School; Bedford College, London; Royal Holloway College, London, B.Sc., D.Sc.
Clarke, Lilian Jane	**39** (1866–1934)	Highbury High School; Royal College of Science, London; University College, London, B.Sc., D.Sc.
Embleton, Alice Laura	**28** (1877–1960)	Sutton High School; University of Wales, Cardiff, B.Sc.
Frankland, Grace Coleridge, Mrs (née Toynbee)	**47** (1858–1946)	Bedford College, London
Fraser, Helen Charlotte Isabella, after 1911 Mrs. Gwynne Vaughan	**26** (1879–1967)	King's College, London; Birkbeck College, London, D.Sc.
Latham, Vida (Viola) Annette	**39** (1866–1958)	Norwich High School, Ellerslie Ladies College; Newnham College, Cambridge; University of London, M.Sc.; University of Michigan, DDS; Northwestern University, Illinois, M.D.

P. Ayres, *Women and the Natural Sciences in Edwardian Britain*,
Palgrave Studies in the History of Science and Technology,
https://doi.org/10.1007/978-3-030-46600-8

Name (unmarried in 1905 unless shown)	*Age (±1) in 1905*	*Education*
Lister, Gulielma	**45** (1860–1949)	Bedford College, London
Ogilvie-Gordon, Maria Matilda, Mrs (née Ogilvie)	**41** (1864–1939)	Boarding school in Edinburgh; Heriot-Watt College, Edinburgh; University College, London; University of Munich, Ph.D.; University of London, D.Sc.
Pertz, Dorothea Frances Matilda	**37** (1859–1939)	Newnham College, Cambridge
Sargant, Ethel	**42** (1863–1918)	North London Collegiate School; Girton College, Cambridge
Saunders, Edith Rebecca	**40** (1865–1945)	Newnham College, Cambridge
Scott, Henderina, Mrs (née Klaassen)	**43** (1862–1929)	Royal College of Science, London
Smith, Annie Lorraine	**51** (1854–1937)	Royal College of Science, London
Veley, Lilian Jane, Mrs (née Nutcombe Gould)	**33** (1872–1936)	Somerville College, Oxford; Trinity College, Dublin, D. Sc.

1(B) The others

Name	*Age in 1905*	*Education*
Duchess of Bedford, Mary Russell	**40** (1865–1937)	Cheltenham Ladies College
Busk, Marian (Lady) (née Balfour)	**45** (1860–1941)	Queen's College, London
Crisp, Catherine, Mrs (née Howes)	**59** (1846–1931)	Not known
Gibbs, Lilian Suzette	**35** (1870–1925)	Swanley College, Kent; Royal College of Science, London
Noel, Amelia Frances	**37** (1868–1950)	Somerville College, Oxford; Swanley College, Kent
Silver, Sarah Marianne, later Mrs Sinclair	**26** (1879–1920)	Not known
Sladen, Constance, Mrs (née Anderson)	**57** (1848–1906)	Not known
Stebbing, Mary Anne, Mrs (née Saunders)	**60** (1845–1927)	Not known
Turner, Emma Louise	**39** (1866–1940)	Private schools
Willmott, Ellen Anne	**47** (1858–1934)	Catholic Convent, Isleworth, Middx.

Bibliography

General Sources

Crawford, Elizabeth. 1999a. *The Women's Suffrage Movement. A Reference Guide 1866–1928*. London: UCL Press.

Creese, Mary. 1998a. *Ladies in the Laboratory? American and British Women in Science, 1800–1900*. London: Scarecrow Press.

Desmond, Ray. 1994a. *Dictionary of British and Irish Botanists and Horticulturalists Including Plant Collectors Flower Painters and Garden Designers*. London: Taylor & Francis.

Haines, Catherine M.C. 2001a. *International Women of Science: A Biographical Dictionary to 1950*. Oxford: ABC-CLIO.

Ogilvie, Marilyn Bailey. 1986. *Women in Science. Antiquity Through the Nineteenth Century. A Biographical Dictionary*. Cambridge, MA: MIT Press.

Oxford Dictionary of National Biography. Oxford: University Press.

References

Allen, D.E. 1976. *The Naturalist in Britain. A Social History*. Princeton: University Press.

———. 1994. Oxford and Druce. *Botanical Society of the British Isles News* 67: 41–45.

Anonymous. 1881. *Essex Field Club Journal* 1: 24–27.

———. 1882a. *Essex Field Club Journal* 2, Appendix, p. xx.

P. Ayres, *Women and the Natural Sciences in Edwardian Britain*, Palgrave Studies in the History of Science and Technology, https://doi.org/10.1007/978-3-030-46600-8

———. 1882b. *Essex Field Club Journal* 2: 62.

———. 1884a. *Essex Field Club Journal* 4: clxxxvi.

———. 1884b. *Essex Field Club Journal* 4: 194.

———. 1894. *Essex Naturalist* 8: 53.

———. 1896a. *Transactions of the British Mycological Society* 1: 3–5.

———. 1896b. *Essex Naturalist* 10: 21.

———. 1899a. Notes. *Nature*, 26 October, 60, 621–625.

———. 1899b. *Nature*, 26 October.

———. 1900. Notes. *Nature* 62: 380.

———. 1901. *British Association for the Advancement of Science, Annual Report, cvi.*

———. 1903. *Essex Naturalist* 13: 182.

———. 1904. *Yearbook of the Royal Society*. London: Harrison and Sons.

———. 1904–1905. *Proceedings of the Linnean Society* 117: 10–11.

———. 1906. The Essex Field Club: What It Is: What It Has Done 1880–1906; and What It Wishes to Do. *Special Publication of the Essex Field Club* 2.

———. 1906–1907. *Proceedings of the Linnean Society* 119: 63.

———. 1907. *Report of the Seventy-Sixth Meeting of the British Association for the Advancement of Science, York, 1906*. London: John Murray.

———. 1908. *Essex Field Club: Yearbook and Calendar for 1908.*

———. 1909. *Country Life*, 29 May, 814.

———. 1910. Obituary. Hendericus M Klaassen FGS. *Geological Magazine*, 191.

———. 1917. Obituary. Thomas McKenny Hughes MA. *Geological Magazine* 4: 334–335.

———. 1927a. *An Outline of the History of the Botanical Department of University College, London*, 11–12. London: University Press, issued by the Department on the occasion of the Centenary of the College.

———. 1927b. *Lady Magazine*, 17 March.

———. 1951. Obituary Notices. Gulielma Lister. *Essex Naturalist* 28: 214.

———. 1969. Schools: Queen's College, Harley Street. In *A History of the County of Middlesex: Volume 1, Physique, Archaeology, Domesday, Ecclesiastical Organization, the Jews, Religious Houses, Education of Working Classes To 1870, Private Education From Sixteenth Century*, ed. J.S. Cockburn, H.P.F. King, and K.G.T. McDonnell. Oxford: University Press for the Institute of Historical Research.

Arber, Agnes. 1903. *The Work of Women in Botanical Research. Biennial leaflet.* London: The Women's Agricultural and Horticultural International Union. Cited by Schmid, Rudolf. 2001. Agnes.

———. 1919. Obituary. Ethel Sargant. *New Phytologist* 18: 120–128.

———. 1927. Ethel Sargant. *Girton Review*, Michaelmas 1927, 17–26.

———. 1939. Miss Dorothea FM Pertz. *Nature* 143: 590–591.

Armstrong, P. 2000. *The English Parson-Naturalist. A Companionship between Science and Religion*. Leominster: Gracewing.

Ayres, P.G. 2005. *Harry Marshall Ward and the Fungal Thread of Death*. St Paul, MN: American Phytopathological Society.

———. 2008. *The Aliveness of Plants. The Darwins at the Dawn of Plant Science*. London: Pickering & Chatto.

———. 2012. *Shaping Ecology. The Life of Arthur Tansley*. Oxford: Wiley-Blackwell.

Bailey, Marilyn Ogilvie. 2000. Obligatory Amateurs: Annie Maunder (1868–1947) and British Women Astronomers at the Dawn of Professional Astronomy. *British Journal for the History of Science* 33: 67–84.

Bailey, Mandy. 2016. Women and the RAS: 100 years of Fellowship. *Astronomy and Geophysics* 57: 19–21.

Baldwin, Melinda. 2015. *Making "Nature". The History of a Scientific Journal*. Chicago: University Press.

Balfour, L. 1943. Lady Busk (1861–1941). *Proceedings of the Linnean Society* 154: 271–272.

Barnett, Henrietta. 1918. *Canon Barnett. His Life, Work, and Friends by His Wife*. Vol. I. London: John Murray.

Beale, N., and Elaine Beale. 2011. *Echoes of Ingen Housz. The Long Lost Story of the Genius Who Rescued the Habsburgs from Smallpox and Became the Father of Photosynthesis*. Salisbury: Hobnob Press.

Bellamy, F.A. 1908. *A Historical Account of the Ashmolean Natural History Society of Oxfordshire 1880–1905*. Oxford: The author.

Bernstein, Susan D. 2006. 'Supposed Differences': Lydia Becker and Victorian Women's Participation in the BAAS. In *Repositioning the Victorian Sciences: Shifting Centres in Nineteenth-Century Scientific Thinking*, ed. David Clifford, Elisabeth Wadge, Alex Warwick, and Martin Willis, 85–93. New York: Anthem.

Berridge, Emily. 1905. Two New Species of *Spencerites insignis*. *Annals of Botany* 19: 273–280.

Bisgrove, R. 1990. *The National Trust Book of the English Garden*. London: Viking.

Blackwell, Elsie M. 1936–1937. Obituary. Margaret Jane Benson. *Proceedings of the Linnean Society* 149: 186–189.

———. 1947–1948. Obituary. Emily M Berridge. *Proceedings of the Linnean Society* 160: 68–70.

Blunt, Alison. 1994. *Travel, Gender and Imperialism. Mary Kingsley and West Africa*. London: The Guilford Press.

Boney, D. 1995. The Botanical 'Establishment' Closes Ranks: Fifteen Days in January 1921. *The Linnean* 11: 26–37.

Boulter, M. 2017. *Bloomsbury Scientists. Science and Art in the Wake of Darwin*. London: UCL Press.

Bower, F.O. 1938. *Sixty Years of Botany in Britain (1875–1935). Impressions of an Eye-Witness.* London: Macmillan.

Bowles, E.A. 1914. *My Garden in Spring.* London: TC and EC Jack.

Britten, J. 1905. Book Notes, News, &c. *Journal of Botany, British and Foreign* 43: 37.

———. 1922. Ethel Sarel Gepp (1864–1922). *Journal of Botany, British and Foreign* 60: 193–195.

Brooke, J., and G. Cantor. 1998. *Reconstructing Nature. The Engagement of Science and Religion.* Edinburgh: T & T Clark.

Brown, Jane. 2000. *The Pursuit of Paradise. A Social History of Gardens and Gardening.* London: Harper Collins.

Brown, Megan K. 2011. *A College for Women, or Something Like It: "Bedford College and the Women's Higher Education Movement, 1849–1900".* Dissertations and Theses University of Portland. Paper 209. Available online.

Bruce, Lynn. 2012. *Scottish Settlement Houses from 1886–1934.* Ph.D. thesis, University of Glasgow, 85–93. Available online.

Burek, Cynthia B. 2007. The Role of Women in Geological Higher Education—Bedford College, London (Catherine Raisin) and Newnham College, Cambridge, UK. In *The Role of Women in the History of Geology*, ed. Cynthia B. Burek and Betty Higgs, vol. 281, 9–38. Geological Society, Special Publications.

Burek, Cynthia V. 2009. The First Female Fellows and the Status of Women in the Geological Society of London. *Geological Society, Special Publications* 317: 373–407.

Burek, Cynthia V., and M. Kolbl-Ebert. 2007. The Historical Problems of Travel for Women Undertaking Geological Fieldwork. In *The Role of Women in the History of Geology*, ed. Cynthia V. Burek and Betty Higgs, vol. 281, 115–122. The Geological Society of London Special Publication.

Buxton, Meriel. 2008. *The High-Flying Duchess. Mary Du Caurroy Bedford 1865–1937.* Leicester: Woodperry Books.

Bynum, H., and W.F. Bynum. 2017. *Botanical Sketchbooks.* London: Thames and Hudson.

Calman, W.C. 1927. Obituary Notices of Fellows Deceased. *Proceedings of the Royal Society, B* 101: i–xxxviii.

Calvert, Agnes. 1887. The Lactiferous Tissue in the Stem of *Hevea brasiliensis. Annals of Botany* 1: 75–77.

Calvert, Agnes, and L.A. Boodle. 1887. On Lactiferous Tissue in the Pith of *Manihot glaziovii*, and on the Presence of Nuclei in This Tissue. *Annals of Botany* 1: 55–62.

Carlile, M.J. 2005. Two Influential Mycologists: Helen Gwynne Vaughan (1879–1967) and Lilian Hawker (1908–1991). *The Mycologist* 19: 129–130.

Chapman, A. 2016. The Lady Astronomers of Victorian Britain. *Astronomy and Geophysics* 57: 12–13.

Cock, A., and D.R. Forsdyke. 2008. *Treasure Your Exceptions: The Science and Life of William Bateson*. New York: Springer.

Cotton, A.D. 1934–1935. Obituary. Miss Ellen Willmott. *Proceedings of the Linnean Society* 147: 195–197.

Crawford, Elizabeth. 1999b. *The Women's Suffrage Movement: A Reference Guide 1866–1928*. London: University College Press.

Creedon, Alison. 2002. A Benevolent Tyrant? The Principles and Practices of Henrietta Barnett (1851–1936), Social Reformer and Founder of Hampstead Garden Suburb. *Women's History Review* 11: 231–252.

Creese, Mary. 1998b. *Ladies in the Laboratory? American and British Women in Science, 1800–1900*. London: Scarecrow Press.

Dare, Deirdre, and Melissa Hardie. 2008. *A Passion for Nature. 19th Century Naturalism in the Circle of Charles Alexander Johns*. Penzance: Patten Press.

Darwin, B. 1943. *British Clubs*. London: William Collins.

De Chadarevian, Soraya. 1996. Laboratory Science Versus Country-House Experiments. The Controversy between Julius Sachs and Charles Darwin. *British Journal of the History of Science* 29: 17–41.

Delf, Ellen. 1944–1945. Obituary Notice. Dr EN Miles Thomas. *Proceedings of the Linnean Society* 156: 235–236.

Desmond, A. 1994b. *Huxley: The Devil's Disciple*. London: Michael Joseph.

Desmond, R. 1995. *Kew: The History of the Royal Botanic Gardens*. London: Harvill Press.

Digby, Lettice. 1924–1925. Obituary. Miss Lilian Suzette Gibbs. *Proceedings of the Linnean Society* 137: 72–73.

Douglas, Gina. 2010. Farquharson, Marian Sarah (1846–1912). *Oxford Dictionary of National Biography*. Oxford: University Press.

Dyhouse, Carol. 1981. *Girls Growing Up in Victorian and Edwardian England*. London: Routledge and Keegan Paul.

———. 1995. *No Distinction of Sex? Women in British Universities, 1870–1939*. UCL Press.

Eivor. 1905. En kvinnlig bakteriolog. *Idun* 26: 325–316.

Embleton, Alice L. 1900. III. On the Structure and Affinities of Echiurus unicinctus. *Transactions of the Linnean Society of London* 8: 77–97.

———. 1901. *Goidelia japonica*—A New Entozoic Copepod from Japan, associated with an Infusorian (*Trichodina*). *Journal of the Linnean Society of London (Zoology)* 28: 211–229.

———. 1903. *Cerataphis lataniae*, a peculiar aphid. *Journal of the Linnean Society (Zoology)* 29: 90–107.

Endersby, J. 2008. *Imperial Nature. Joseph Hooker and the Practices of Victorian Science*. Chicago: University Press.

Fara, Patricia. 2018. *A Lab of One's Own. Science and Suffrage in the First World War*. Oxford: Oxford University Press.

Farquharson, Marian S. 1885–1886. Notes on Mosses of the North of Scotland. *Scottish Naturalist* 8: 381.

Farquharson, F.R.M.S. 1886. *On the Identification of the British Mosses by Their Distinctive Characters*. British Association for the Advancement of Science: Held at Aberdeen in September 1885. London: John Murray.

Farquharson, Marian S. 1889–1890. Ferns and Mosses of the Alford District. *Scottish Naturalist* 10: 193–198.

———. 1900. *Women in Professions*. The Professional Section of the International Congress of Women, London, July 1899, 184–185, ed., Lady Ishbel Aberdeen. London: T Fisher Unwin.

Fawcett, Millicent Garrett. 1925. *What I Remember*. New York: GP Putnam's.

Fogg, E. 1934. Obituaries. Dr. Lilian Clarke. *Proceedings of the Linnean Society* 1933–1934: 150–151.

Forrester, J., and Laura Cameron. 2017. *Freud in Cambridge*. Cambridge: University Press.

Frank, Katherine. 2005. *A Voyager Out: The Life of Mary Kingsley*. New York: Tauris Parke Paperbacks.

Frankland, Grace C., and P.F. Frankland. 1887a. Studies of Some New Micro-organisms Obtained from the Air. *Philosophical Transactions of the Royal Society of London, B* 178: 257–287.

———. 1887b. On Some New and Typical Micro-organisms Obtained from Water and Soil. *Proceedings of the Royal Society of London* 43: 414–418.

Fraser-Mackintosh, C. 1898. *An Account of the Confederation of Clan Chattan; Its Kith and Kin*. Glasgow: John Mackay.

Gage, A.T., and W.T. Stearn. 1988. *A Bicentenary History of the Linnean Society of London*. London: Academic Press.

Garner, W.E. 1948. Obituary Notice. Percy Faraday Frankland. 1858–1946. *Journal of the Chemical Society (Resumed)* 3: 1996–2005.

Gaycken, O. 2012. The Secret Life of Plants. Visualising Vegetative Movement, 1880–1903. *Early Popular Visual Culture* 10: 51–69.

Geison, G.L. 1978. *Michael Foster and the Cambridge School of Physiology*. Princeton, NJ: University Press.

Gianquitto, Tina. 2013. Botanical Smuts and Hermaphrodites: Lydia Becker, Darwin's Botany and Educational Reform. *Isis* 104: 250–277.

Glick, Daphne. 1995. *The National Council of Women of Great Britain: The First One Hundred Years 1895–1995*. London: National Council of Women of Great Britain.

Godwin, H. 1957. Arthur George Tansley, 1871–1955. *Biographical Memoirs of Fellows of the Royal Society* 3: 227–246.

———. 1985. *Cambridge and Claire*. Cambridge: University Press.

Goodman, M. 2007. *Suffer and Survive. The Extreme Life of Dr. JS Haldane.* London: Simon & Schuster.

Gore, J. 1938. *Mary Duchess of Bedford 1865–1937*, 2 vols. London: John Murray.

Gould, Paula. 1997. Women and Culture of University Physics in Late Nineteenth-Century Cambridge. *British Journal for the History of Science* 30: 127–149.

Griffin, Penny. 1986. *St Hugh's: One Hundred Years of Women's Education in Oxford.* London: Palgrave Macmillan.

Groves, J. 1913. Henry Groves (1855–1912). *Journal of Botany* 51: 73–79.

Haines, Catherine M.C. 2001b. *International Women of Science: A Biographical Dictionary to 1950.* Oxford: ABC-CLIO.

Hale Bellot, H. 1929. *University College, London, 1826–1926.* London: University Press.

Hall, Ruth. 1977. *Marie Stopes. A Biography.* London: Virago.

Hamer, Emily. 2016. *Britannia's Glory. A History of Twentieth Century Lesbians.* London: Bloomsbury Academic.

Hamshaw Thomas, H. 1947. Obituary. Edith Rebecca Saunders. *Proceedings of the Linnean Society* 158: 75–77.

Haskins, E.P. 1999. Miss Gulielma Lister F.L.S. Remembered. *Mycologist* 13: 54–56.

Heffer, S. 2017. *The Age of Decadence. Britain 1880 to 1914.* London: Random House Books.

Herries Davies, G.L. 2007. *Whatever is Under the Earth. The Geological Society of London 1807 to 2007.* London: The Geographical Society of London.

Higgitt, Rebekah, and C.W.J. Withers. 2008. Science and Sociability. Women as Audience at Meetings of the British Association for the Advancement of Science, 1831–1901. *Isis* 99: 1–27.

Hobhouse, Hermione. 2002. *The Crystal Palace and the Great Exhibition: Art, Science and Productive Industry. A History of the Royal Commission for the Exhibition of 1851.* London: Continuum.

Howell, A.C. 2005. James Lomax (1857–1934): Palaeobotanical Catalyst or Hindrance? In *History of Palaeobotany. Selected Essays*, ed. A.J. Bowden, V. Burek Cynthia, and R. Wilding, 241. Geological Society of London, Special Publications.

Hudson, Briony. 2013. *The School of Pharmacy, University of London: Medicines, Science, and Society, 1842–1912.* London: Academic Press and Elsevier.

Huffman, Joan B. 2018. *Lady Frances. Frances Balfour, Aristrocrat Suffragist.* Kibworth: Matador Press.

Hughes, Mary V. 1978a. *A London Girl of the 1880s.* Oxford: University Press.

———. 1978b. *A London Home in the 1890s.* Oxford: University Press.

Humphry, Charlotte E. 1897. *Manners for Women.* London: Ward, Lock & Co.

———. 1898. *A Word to Women.* London: James Bowden.

Hutchinson, Horace G. 1914. *Life of Sir John Lubbock. Lord Avebury*. Vol. ii. London: Macmillan.

Izzard, Molly. 1969. *A Heroine in Her Time: A Life of Dame Helen Gwynne-Vaughan 1879–1967*. London: Macmillan.

Jackson, M.B. 2015. One Hundred and Twenty-Five Years of the *Annals of Botany*. Part I. The first 50 years (1887–1936). *Annals of Botany* 115: 1–18.

Jekyll, Gertrude, and E. Mawley. 1902. *Roses for English Gardens*, From the Country Life Library. London: George Newnes.

Jones, Claire G. 2009. Bodies of Controversy? Women and the Royal Society. In *Femininity, Mathematics and Science, 1880–1914*, ed. Claire G. Jones, 175–204. Berlin: Springer.

———. 2010. Henderina (Rina) Scott. Botanist and filmmaker. *HerStoria*, Summer, 40–43.

———. 2016. The Tensions of Homemade Science in the Work of Henderina Scott and Hertha Ayrton. In *Domesticity in the Making of Modern Science*, ed. Donald L. Opitz, Staffan Bergwik, and Brigitte Van Tiggelen, 84–104. London: Springer.

Jones, Claire G., and Sue Hawkins. 2015. Women and Science. *Notes and Records of the Royal Society* 69: 5–9.

Keighren, Innes M. 2017. 'A Royal Geographical Society for Ladies': The Lyceum Club and Ladies Geographical Frontiers in Edwardian London. *The Professional Geographer* 69: 661–669.

Lefebvre, T., J. Malthete, and L. Mannoni. 1999. *Letters from Étienne-Jules Marey to Georges Demenÿ 1880–1894*. Paris: AHFRC.

Lewis, D. 1969. The Genetical Society—The First Fifty Years. In *Fifty Years of Genetics. Proceedings of a Symposium held at the 160th Meeting of the Genetical Society on the 50th Anniversary of its Founding*, ed. John Jinks, 1–7. Edinburgh: Oliver and Boyd.

Liévre, Le, and Audrey. 1980. *Miss Willmott of Warley Place. Her Life and Her Gardens*. London: Faber and Faber.

Lister, A. 1906. The Epping Forest Foray: 1st to the 6th October 1906. *Transactions of the British Mycological Society* 2: 133–141.

Lister, G. 1927. On a Collection of Plants Made by the Lord Lister between the Years 1844–1848, Many of Them from the Neighbourhood of Upton. *Essex Naturalist* 21: 104–107.

Locke, Audrey. 1916. *The Hanbury Family*, 2 vols. London: Arthur L Humphreys.

Lockyer, N. 1869. Lectures to Ladies. *Nature* 1: 45–46.

———. 1870. The Scientific Education of Women. *Nature* 2: 117–118.

MacPherson, W.M. 1895. *Materials for the History of a Church and Priory of Monymusk*. Aberdeen: Taylor and Henderson.

Marner, Serena. 2001. *In Celebration of the Centenary of the Ashmolean Natural History Society of Oxfordshire* 1901–2001. pdf online

———. 2002. *In Celebration of the Centenary of the Ashmolean Natural History Society of Oxfordshire 1901–2001.* The Ashmolean Natural History Society of Oxfordshire, 8pp. Available online.

Marsden, J. 2003. Charter and Bye-Laws. *The Linnean* 19: 13–15.

Mason, Joan. 1991. Hertha Ayrton (1854–1923) and the Admission of Women to the Royal Society of London. *Notes and Records of the Royal Society of London* 45: 201–220.

———. 1995a. The Women Fellows' Jubilee. *Notes and Records of the Royal Society of London* 49: 126.

———. 1995b. The Women Fellows' Jubilee. *Notes and Records of the Royal Society* 49: 125–140.

———. 2004. Ayrton [née Marks] Sarah [Phoebe] Hertha (1854–1923). *Oxford Dictionary of National Biography* online https://doi.org/10.1093/ref:odnb/37136.

McDonald, Deborah. 2004. *Clara Collet 1860–1948. An Educated Working Woman.* London: Woburn Press.

Meadows, A.J. 2008. *Science and Controversy. A Biography of Sir Norman Lockyer, Founder of Nature.* London: Palgrave Macmillan.

Meldola, R., and W. White. 1885. *Report on the East Anglian Earthquake of April 22nd, 1884*, Essex Field Club Special Memoirs, 1. London: Macmillan.

Metcalfe, C.R. 1942. A Short History of the Jodrell Laboratory. *Chronica Botanica* 7: 174–176.

Moore, Charlotte. 2010. *Hancox. A House and a Family.* London: Viking for Penguin Books.

Moore, J.E.S., and Alice L. Embleton. 1906. On the Synapsis in Amphibia. *Proceedings of the Royal Society, B* 77: 555–562.

Morris, M.C.F. 1897. *Francis Orpen Morris. A Memoir.* London: John Nimmo.

Morton, A.G., and Mary Noble. 1983. Botany and Mycology. *Proceedings of the Royal Society of Edinburgh, B* 84: 65–83.

Nichols, D. 2003. A Biography of Percy Sladen (1849–1900). *The Linnean* (special issue no. 4): 30pp.

O'Connor, W.J. 1991. *British Physiologists, 1885–1914. A Biographical Dictionary.* Manchester: University Press.

Ogilvie-Gordon, Maria M. 1897. Microscopic and Systematic Study of Madreporian Fossils. *Philosophical Transactions of the Royal Society* 187B: 83–345.

Oliver, F.W. 1906. The Bouche d'Erquy in 1906. *New Phytologist* 5: 189–195.

———. 1933. *Letters from Egypt. Mildred Alice Oliver: 1 January 1869–8 October 1932.* London: Euston Press for private circulation. (Sackler Library, University of Oxford).

———. 1935. Dukinfield Henry Scott, 1954–1934. *Annals of Botany* XLIX: 823–837.

Oliver, F.W., and A.C. Seward. 1934. Dukinfield Henry Scott 1854–1934. *Obituary Notices of Fellows of the Royal Society* 1: 205–227.

Opitz, D.L. 2004. 'Behind Folding Shutters in Whittingehame House': Alice Blanche Balfour (1850–1936) and Amateur Natural History. *Archives of Natural History* 31: 330–384.

———. 2014. 'Back to the Land': Lady Warwick and the Movement for Women's Collegiate Agricultural Education. *Agricultural History Review* 62: 119–145.

Parry, J. 2011. Broadland's Bittern Pioneer. *Tern* (Norfolk Wildlife Trust), Summer ed.

Parry, J., and J. Greenwood 2020. *Emma Turner. A Life Looking at Birds*. Norwich: Norfolk and Norwich Naturalists' Society.

Pearson, H.L. 2005. Marie Stopes and a Century of Palaeobotany. *Linnean* 21: 25–28.

Pedersen, Sarah. 2004. *Within Their Sphere? Correspondence to the Aberdeen Daily Newspapers, 1900–1918*. Ph.D. thesis, The Robert Gordon University, Aberdeen.

Pim, Sheila. 1966. *The Wood and the Trees. A Biography of Augustine Henry*. London: Macdonald.

Poulton, E.B. 1890. *The Colours of Animals*. London: Kegan Paul, Trench and Tubner.

———. 1899–1900. The Corresponding Societies Committee of the British Association, Bradford, 1900. *Essex Naturalist* 11: 310–311.

———. 1919. *The Life of Ronald Poulton*. London: Sidgwick & Johnson.

Pugsley, H.W. 1933. Obituary. James Groves (1858–1933). *Journal of Botany* 71: 136–139.

Purvis, June. 1991. *A History of Women's Education in England*. Milton Keynes: Open University Press.

Quinn, Susan. 1996. *Marie Curie. A Life*. London: Mandarin.

Raikes, Elizabeth. 1908. *Dorothea Beale of Cheltenham*. London: Archibald Constable.

Rappaport, Helen. 2001. *Encyclopedia of Women Social Reformers*. Frances Mary Buss. Oxford: ABC-Clio.

Raverat, Gwen. 1954. *Period Piece. A Cambridge Childhood*. London: Faber and Faber.

Rayner-Canham, Marelene F., and G.W. Rayner-Canham. 2003. Pounding on the Doors: The Fight for Acceptance of British Women Chemists. *Bulletin of the History of Chemistry* 28: 110–119.

———. 2008. *Chemistry was Their Life. Pioneering British Women Chemists, 1880–1949*. London: Imperial College Press.

Reynolds Green, J. 1909. *A History of Botany 1860–1900*. Oxford: Clarendon Press.

Richmond, Marsha L. 1997. 'A Lab of One's Own'. The Balfour Biological Laboratory for Women at Cambridge University, 1884–1914. *Isis* 88: 422–455.

———. 2006. The 'Domestication' of Heredity: The Familial Organisation of Geneticists at Cambridge University, 1895–1910. *Journal of the History of Biology* 39: 565–605.

Ridley, Marian S. 1881. *A Pocket Guide to British Ferns.* London: David Bogue.

Robinson, E. 2007. The Influential Muriel Arber: A Personal Reflection. In *The Role of Women in the History of Geology*, ed. Cynthia V. Burek and Bettie Higgs, 287–294. London: Geological Society.

Robinson, S., and Kate Perry. 2008. Science at Girton—Ethel Sargant. *Girton College Annual Review*: 21–23.

Roy, J. 1890. Freshwater Algae of Enbridge Lake and Vicinity, Hampshire. *Journal of Botany* 28: 334–338.

Royle, Alice C. 1903. Mrs Farquharson of Haughton FRMS. Botanist and Writer. *Boudoir* 2: 811–816.

Ruskin, J. 1865. *Sesame and Lilies.* London: John Wiley.

———. 1886. *Proserpina, Vol. 1, Studies of Wayside Flowers.* Orpington: George Allen.

Russell, C.A. 1996. *Edward Frankland. Chemistry, Controversy and Conspiracy in Victorian England.* Cambridge: University Press.

Sanders, Dawn. 2005. From Radicle to Radical: Biology Education and the First Women Fellows of the Linnean Society of London. *The Linnean* 21: 20–24.

———. 2007. *The Global and the Local: The History of Science and the Cultural Integration of Europe.* Proceedings of the 2nd ICESHS (Cracow, Poland, September 6–9, 2006), ed. M. Kokowski, 606–607. Krakow: Polish Academy of Arts and Sciences.

Sargant, Ethel. 1900. Women and Original Research. *Frances Mary Buss Schools Jubilee Magazine*, November.

———. 1901. The Inheritance of a University. *Girton Review*: 9–21.

Savage, S. 1936–1937. Obituaries. Mary du Caurroy Russell, Her Grace the Duchess of Bedford. *Proceedings of the The Linnean Society* 149: 185.

Schiebinger, Londa. 1999. *Has Feminism Changed Science?* Cambridge, MA: Harvard University Press.

Schmid, Rudolf. 2001. Agnes Arber, née Robertson (1879–1960): Fragments of Her Life, Including Her Place in Biology and in Women's Studies. *Annals of Botany* 88: 1105–1128.

Scott, D.H. 1887. On Nuclei in *Oscillaria* and *Tolypothrix. Journal of the Linnean Society, Botany* 24: 188–192.

Scott, Rina. 1903. On the Movements of the Flowers of *Sparmannia Africana*, and Their Demonstration by Means of Kinematograph. *Annals of Botany* 17: 761–777.

Scott, Dukinfield H. 1907. Animated Photographs of Plants. *Journal of the Royal Horticultural Society* 32: 48–51.

Scott, D.H. 1908. Arthur Lister F.R.S. *Journal of Botany, British and Foreign* xlvi: 331–334.

———. 1925. German Reminiscences of the Early 'Eighties. *New Phytologist* 24: 9–16.

Scott, Caroline. 2017. *Holding the Home Front. The Women's Land Army in the First World War*. Barnsley: Pen & Sword Press.

Seward, A.C. 1934. Dukinfield Henry Scott, 1854–1934. *Obituary Notices of Fellows of the Royal Society* 1: 205–227.

Sheffield, Suzanne le-May. 2006. Gendered Collaborations: Marrying art and Science. In *Figuring It Out: Science, Gender and Visual Culture*, ed. Ann Shteir and Bernard Lightman, 240–266. London: Dartmouth College Press.

Sidgwick, Eleanor. 1897. *University Education of Women. A Lecture Delivered at University College, Liverpool, May 1896*, 26pp. Cambridge: Macmillan & Bowes.

Small, E. 2013. *Mary Lily Walker. Forgotten Visionary of Dundee*. Dundee: University Press.

Stearn, W.R. 1960. Mrs Agnes Arber: Botanist and Philosopher, 1879–1960. *Taxon* 9: 161–263.

Stirling, A.M.W. 1922. *William De Morgan and His Wife*. New York: Henry Holt and Co.

Stopes, Marie. 1903. On the Leaf Structure of '*Cordaites*'. *New Phytologist* 2: 92–98.

Sutherland, Gill. 2006. *Faith, Duty and the Power of Mind. The Cloughs and Their Circle, 1820–1960*. Cambridge: University Press.

Swanwick, Helena M. 1935. *I Have Been Young*. London: Victor Gollancz.

Tansey, Tilli. 2015. Women and the Early Journal of Physiology. *Journal of Physiology* 593: 347–350.

Thompson, P. 1930. A Short History of the Essex Field Club. *Essex Field Cub Special Memoirs* 7: 2.

Throesch, Elizabeth L. 2017. *Before Einstein: The Fourth Dimension Fin-de-Siecle Literature and Culture*. London: Athena Press.

Toynbee, Gertrude. 1909. *Joseph Toynbee FRS. Aural Surgeon*. London: Henry J. Glaister.

Tuke, Margaret J. 1939. *A History of Bedford College for Women 1849–1939*. Oxford: University Press.

Uglow, Jenny. 2004. *A Little History of British Gardening*. London: Chatto and Windus.

Van der Pas, P.W. 1970. The Correspondence of Hugo de Vries and Charles Darwin. *Janus* 57: 173–213.

Veley, Lilian J. 1902a. Birds Capturing Butterflies in flight. *Nature* 65: 299.

———. 1902b. Birds Attacking Butterflies and Moths. *Nature* 65: 392.

———. 1905. A further contribution to the study of Pelomyxa palustris (Greeff). *Journal of the Linnean Society* (Zoology) 29: 374–95.

Veley, V.H., and Lilian J. Veley. 1898. *The Microorganisms of Faulty Rum*. London: Henry Frowde.

Wachtler, M., and Cynthia V. Burek. 2007. Maria Matilda Ogilvie Gordon (1864–1939): A Scottish Researcher in the Alps. In *The Role of Women in the History of Geology*, ed. Cynthia V. Burek and Bettie Higgs, 305–317. London: Geological Society Special Publications, No. 281.

Walker, Lynne. 2006. Women Patron-Builders in Britain: Identity, Difference and Memory in Spatial and Material Culture. In *Local/Global. Women Artists in the Nineteenth Century*, ed. Deborah Cherry and Janice Helland, 121–136. Ashgate: Colchester.

Walker, C.E., and Alice L. Embleton. 1906. On the Origin of the Sertoli or Foot-Cells of the Testis. *Proceedings of the Royal Society, B* 78: 50–52.

———. 1908. Observations on the Nucleoli in the Cells of *Hydra fusca*. *Quarterly Journal of Experimental Physiology* 1: 287–290.

Warburton, C., and Embleton, Alice L. 1902. The life history of the black-currant gall-mite, Eriophyses (Phytoptis) ribis. Westwood. *Journal of the Linnean Society (Zoology)* 28: 366–378.

Watts, Ruth. 2007. *Women in Science. A Social and Cultural History*. London: Routledge.

Way, Twigs. 2009. *Garden Gnomes: A History*. London: Bloomsbury Publishing.

Whiting, M. Muriel. 1949–1950. Obituaries. Miss Emilia Frances Noel. *Proceedings of the Linnean Society* 162: 232.

Williamson, W.C. 1896. *Reminiscences of a Yorkshire Naturalist*. Edited by his Wife. London: George Redway.

Index[1]

[1] Note: Page numbers followed by 'n' refer to notes.

P. Ayres, *Women and the Natural Sciences in Edwardian Britain*, Palgrave Studies in the History of Science and Technology, https://doi.org/10.1007/978-3-030-46600-8

R

S

9783030465995